Preface

Primary Mathematics (U.S. Edition) is a complete program based on
Mathematics series from Singapore. Designed to equip students with a
topics are covered in depth and taught to mastery. By focusing on mathematical understanding, the program aims to help students develop logical thinking and problem solving skills.

The Core Curriculum for each semester comprises one Textbook and one Workbook. The textbook is a non-consumable class activity book and is used during class instruction. The workbook is consumable and is used for independent work.

The **Primary Mathematics (U.S. Edition)** program calls for direct instruction and focuses on mathematical thinking with immediate application of new skills to problem solving. By encouraging students to solve problems in a variety of ways, this program promotes an understanding of the way mathematical processes work.

The **Primary Mathematics (U.S. Edition)** program follows a Concrete → Pictorial → Abstract approach. This enables students to encounter math in a meaningful way and translate mathematical skills from the concrete to the abstract. It allows students to understand mathematical concepts before learning the rules of formulaic expressions. Students first encounter the mathematical concepts through the use of manipulatives. They then move on to the pictorial stage in which pictures are used to model problems. When they are familiar with the ideas taught, they progress to a more advanced or abstract stage in which only numbers, notation, and symbols are used.

The **Primary Mathematics Teacher's Guide** is designed for teachers new to the program. It provides background notes on the mathematical concepts for each unit and relates the concepts being learned in a specific unit to the program as a whole. It includes suggestions for presenting the concepts concretely and for differentiating instruction. It also provides detailed lesson plans and some additional suggested problems. Answers are provided for all problems in both the textbook and workbook, and often include suggested solutions (not all possible methods of solution are given). The appendix includes Mental Math worksheets and other aids. It is hoped that this guide will increase the teacher's understanding of mathematics in general, as well as how it is taught in this program.

The **Primary Mathematics Teacher's Guide** is a *guide*. Although the suggested lessons are detailed, each classroom is unique, and each teacher is unique. One guide cannot anticipate all possible situations. Experienced teachers can and should bring their own methods, experiences, and strengths to the classroom in teaching the concepts to the students.

It is important that students understand concepts and not just follow procedures, but they still must have adequate practice in following the standard procedures correctly, such as the addition and subtraction algorithms. Following a procedure alone is not sufficient if the underlying concepts are not understood. By requiring students to apply their knowledge in new situations, you can determine whether they understand the concepts. It is also important that students learn to reason through word problems and solve them logically, rather than being required to follow a step-by-step procedure for problem solving that can only apply to problems that lend themselves to those specific steps. Otherwise, students will not learn to reason through problems that do not work well with a given predetermined set of steps. Strategies that can apply to many types of problems are more valuable than strategies that apply to only a few types of problems, or only to easier problems.

Allow students sufficient use of manipulatives so that they are seen as tools, rather than toys.

Table of Contents

Developmental Continuum

Primary Mathematics 1A

- 1 Numbers 0 to 10
  - 1 Counting
- 2 Number Bonds
  - 1 Making Number Stories
- 3 Addition
  - 1 Making Addition Stories
  - 2 Addition With Number Bonds
  - 3 Other Methods of Addition
- 4 Subtraction
  - 1 Making Subtraction Stories
  - 2 Methods of Subtraction
- 5 Ordinal Numbers
  - 1 Naming Position
- 6 Numbers to 20
  - 1 Counting and Comparing
  - 2 Addition and Subtraction
- 7 Shapes
  - 1 Common Shapes
- 8 Length
  - 1 Comparing Length
  - 2 Measuring Length
- 9 Weight
  - 1 Comparing Weight
  - 2 Measuring Weight

Primary Mathematics 2A

- 1 Numbers to 1000
  - 1 Looking Back
  - 2 Comparing Numbers
  - 3 Hundreds, Tens and Ones
- 2 Addition and Subtraction
  - 1 Meanings of Addition and Subtraction
  - 2 Addition Without Renaming
  - 3 Subtraction Without Renaming
  - 4 Addition With Renaming
  - 5 Subtraction With Renaming
- 3 Length
  - 1 Measuring Length in Meters
  - 2 Measuring Length in Centimeters
  - 3 Measuring Length in Yards and Feet
  - 4 Measuring Length in Inches
- 4 Weight
  - 1 Measuring Weight in Kilograms
  - 2 Measuring Weight in Grams
  - 3 Measuring Weight in Pounds
  - 4 Measuring Weight in Ounces
- 5 Multiplication and Division
  - 1 Multiplication
  - 2 Division
- 6 Multiplication Tables of 2 and 3
  - 1 Multiplication Table of 2
  - 2 Multiplication Table of 3
  - 3 Dividing by 2
  - 4 Dividing by 3

Primary Mathematics 3A

- 1 Numbers to 10,000
  - 1 Thousands, Hundreds, Tens and Ones
  - 2 Number Patterns
- 2 Addition and Subtraction
  - 1 Sum and Difference
  - 2 Adding Ones, Tens, Hundreds and Thousands
  - 3 Subtracting Ones, Tens, Hundreds and Thousands
  - 4 Two-Step Word Problems
- 3 Multiplication and Division
  - 1 Looking Back
  - 2 More Word Problems
  - 3 Multiplying Ones, Tens and Hundreds
  - 4 Quotient and Remainder
  - 5 Dividing Hundreds, Tens and Ones
- 4 Multiplication Tables of 6, 7, 8 and 9
  - 1 Looking Back
  - 2 Multiplying and Dividing by 6
  - 3 Multiplying and Dividing by 7
  - 4 Multiplying and Dividing by 8
  - 5 Multiplying and Dividing by 9
- 5 Money
  - 1 Dollars and Cents
  - 2 Addition
  - 3 Subtraction

Primary Mathematics 1B

- 1 Comparing Numbers
  - 1 Comparing Numbers
  - 2 Comparison by Subtraction
- 2 Graphs
  - 1 Picture Graphs
- 3 Numbers to 40
  - 1 Counting
  - 2 Tens and Ones
  - 3 Addition and Subtraction
  - 4 Adding Three Numbers
- 4 Multiplication
  - 1 Adding Equal Groups
  - 2 Making Multiplication Stories
  - 3 Multiplication Within 40
- 5 Division
  - 1 Sharing and Grouping
- 6 Halves and Quarters
  - 1 Making Halves and Quarters
- 7 Time
  - 1 Telling Time
- 8 Numbers to 100
  - 1 Tens and Ones
  - 2 Order of Numbers
  - 3 Addition Within 100
  - 4 Subtraction Within 100
- 9 Money
  - 1 Bills and Coins
  - 2 Shopping

Primary Mathematics 2B

- 1 Addition and Subtraction
  - 1 Finding the Missing Number
  - 2 Methods for Mental Addition
  - 3 Methods for Mental Subtraction
- 2 Multiplication and Division
  - 1 Multiplying and Dividing by 4
  - 2 Multiplying and Dividing by 5
  - 3 Multiplying and Dividing by 10
- 3 Money
  - 1 Dollars and Cents
  - 2 Adding Money
  - 3 Subtracting Money
- 4 Fractions
  - 1 Halves and Quarters
  - 2 Writing Fractions
- 5 Time
  - 1 Telling Time
  - 2 Time Intervals
- 6 Capacity
  - 1 Comparing Capacity
  - 2 Liters
  - 3 Gallons, Quarts, Pints and Cups
- 7 Graphs
  - 1 Picture Graphs
- 8 Geometry
  - 1 Flat and Curved Faces
  - 2 Making Shapes
- 9 Area
  - 1 Square Units

Primary Mathematics 3B

- 1 Mental Calculation
  - 1 Addition
  - 2 Subtraction
  - 3 Multiplication
  - 4 Division
- 2 Length
  - 1 Meters and Centimeters
  - 2 Kilometers
  - 3 Yards, Feet and Inches
  - 4 Miles
- 3 Weight
  - 1 Kilograms and Grams
  - 2 More Word Problems
  - 3 Pounds and Ounces
- 4 Capacity
  - 1 Liters and Milliliters
  - 2 Gallons, Quarts, Pints and Cups
- 5 Graphs
  - 1 Bar Graphs
- 6 Fractions
  - 1 Fraction of a Whole
  - 2 Equivalent Fractions
- 7 Time
  - 1 Hours and Minutes
  - 2 Other Units of Time
- 8 Geometry
  - 1 Angles
  - 2 Right Angles
- 9 Area and Perimeter
  - 1 Area
  - 2 Perimeter
  - 3 Area of a Rectangle

Primary Mathematics 4A

1 Whole Numbers
  1 Numbers to 100,000
  2 Rounding off Numbers
  3 Factors
  4 Multiples
2 Multiplication and Division of Whole Numbers
  1 Multiplication by a 1-digit Number, Division by a 1-digit Number and by 10
  2 Multiplication by a 2-digit Number
3 Fractions
  1 Adding Fractions
  2 Subtracting Fractions
  3 Mixed Numbers
  4 Improper Fractions
  5 Fraction of a Set
4 Tables and Graphs
  1 Presenting Data
5 Angles
  1 Measuring Angles
6 Perpendicular and Parallel Lines
  1 Perpendicular Lines
  2 Parallel Lines
7 Area and Perimeter
  1 Rectangles and Squares
  2 Composite Figures

Primary Mathematics 5A

1 Whole Numbers
  1 Place Values
  2 Millions
  3 Approximation and Estimation
  4 Multiplying by Tens, Hundreds or Thousands
  5 Dividing by Tens, Hundreds or Thousands
  6 Order of Operations
  7 Word Problems
2 Multiplication and Division by a 2-digit Whole Number
  1 Multiplication
  2 Division
3 Fractions
  1 Fraction and Division
  2 Addition and Subtraction of Unlike Fractions
  3 Addition and Subtraction of Mixed Numbers
  4 Product of a Fraction and a Whole Number
  5 Product of Fractions
  6 Dividing a Fraction by a Whole Number
  7 Word Problems
4 Area of Triangle
  1 Finding the Area of a Triangle
5 Ratio
  1 Finding Ratio
  2 Equivalent Ratios
  3 Comparing Three Quantities
6 Angles
  1 Measuring Angles
  2 Finding Unknown Angles

Primary Mathematics 6A

1 Algebra
  1 Algebraic Expressions
2 Solid Figures
  1 Drawing Solid Figures
  2 Nets
3 Ratio
  1 Ratio and Fraction
  2 Ratio and Proportion
  3 Changing Ratios
4 Percentage
  1 Part of a Whole as a Percentage
  2 One Quantity as a Percentage of Another
  3 Solving Percentage Problems by Unitary Method
5 Speed
  1 Speed and Average Speed

Primary Mathematics 4B

1 Decimals
  1 Tenths
  2 Hundredths
  3 Thousandths
  4 Rounding off
2 The Four Operations of Decimals
  1 Addition and Subtraction
  2 Multiplication
  3 Division
3 Measures
  1 Multiplication
  2 Division
4 Symmetry
  1 Symmetric Figures
5 Solid Figures
  1 Identifying Solid Figures
6 Volume
  1 Cubic Units
  2 Volume of a Cuboid

Primary Mathematics 5B

1 Decimals
  1 Approximation and Estimation
  2 Multiplication by Tens, Hundreds or Thousands
  3 Division by Tens, Hundreds or Thousands
  4 Multiplying by a 2-digit Whole Number
  5 Conversion of Measurements
2 Percentage
  1 Percent
  2 Writing Fractions as Percentage
  3 Percentage of a Quantity
3 Average
  1 Average
4 Rate
  1 Rate
5 Graphs
  1 Line Graphs
6 Triangles
  1 Sum of Angles of a Triangle
  2 Isosceles and Equilateral Triangles
  3 Drawing Triangles
7 4-sided Figures
  1 Parallelograms, Rhombuses and Trapezoids
  2 Drawing Parallelograms and Rhombuses
8 Tessellations
  1 Tiling Patterns
9 Volume
  1 Cubes and Cuboids
  2 Finding the Volume of a Solid

Primary Mathematics 6B

1 Fractions
  1 Division
  2 Order of Operations
  3 Word Problems
2 Circles
  1 Radius and Diameter
  2 Circumference
  3 Area
3 Graphs
  1 Pie Charts
4 Volume
  1 Solving Problems
5 Triangles and 4-sided Figures
  1 Finding Unknown Angles
6 More Challenging Word Problems
  1 Whole Numbers and Decimals
  2 Fractions
  3 Ratio
  4 Percentage
  5 Speed

Material

It is important to introduce the concepts concretely, but it is not important exactly what manipulative is used. Teachers need to use equivalent material that can be displayed on the board, either by using an overhead projector, or objects with magnetic or sticky backs.

Whiteboard and dry-erase markers
An individual whiteboard for each student allows them to work out the problems presented during a lesson. Each student can then hold up the board when he or she has finished the problem.

Counters
Round counters are easy to use and pick up, but any type of counter will work.

Hundred-chart
A number chart from 1-100. Laminated ones for students are useful, since they can be used with dry-erase markers. There is a hundred-chart in the appendix that can be copied.

Multilink cubes
These are cubes that can be linked together on all 6 sides. At this level, it is sufficient that the cubes can be linked together on only two sides, however, cubes that can be linked on more than two sides are useful for creating arrays.

Base-10 set
A set usually has 100 unit-cubes, 10 or more ten-rods, 10 hundred-flats, and 1 thousand-block.

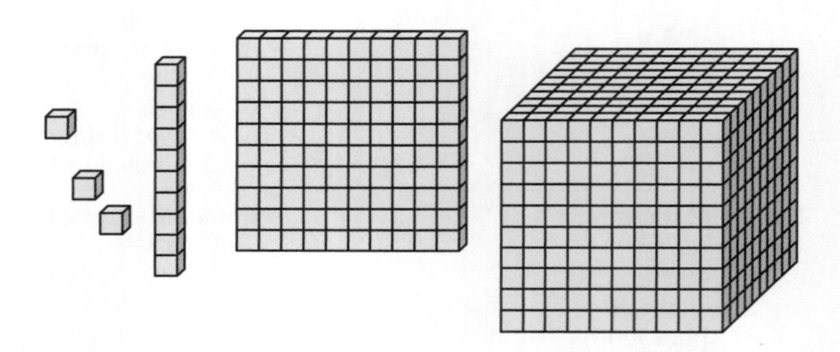

Place-value discs
Round discs with 1, 10, or 100 written on them. You can label round counters using a permanent marker. You need 20 of each kind for each student.

Place-value chart
You can draw a simple one on paper or a whiteboard. It should be large enough to use with number discs or base-10 blocks. You can draw one and put it along with some cardboard backing into a sheet protector, or they can be laminated. That way, students can use dry-erase markers to write numbers or draw number discs on the chart.

Hundreds	Tens	Ones
100 100 100	10 10 10 10 10 10	1 1 1 1

Place-value cards
Place-value cards show the value of each digit and can be fitted on top of each other to form a numeral. Copy or cut out the ones in the appendix, or make your own with index cards, or buy them. They are only used for a few lessons at the beginning and are beneficial primarily for struggling students.

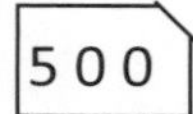

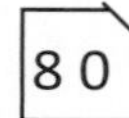

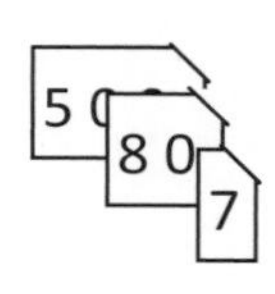
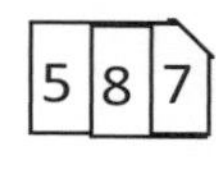

Math fact cards
Cards with math expressions for addition and subtraction within 20 and multiplication and division for 2 and 3. Answers should be on separate cards for games. You can use index cards. If you laminate them, they will last longer.

Number cubes
Cubes that you can label and throw, like dice. These are for group games. You need two per group.

Number cards
For games, cards with 0, 1, 2, 3, 4, 5, 6, 7, 8, 9, or 10 written on them. You need four sets per group for games. You will need various other number cards, which can be made from index cards. See individual lessons.

Measuring tools
Meter sticks and yard sticks, 1 per group of students.
Rulers (12 inch/30 cm) for each student.
Measuring tape marked in inches and centimeters, 1 per group of students.

Supplementary math
Singapore Math Inc®carries a number of supplementary workbooks that can be used for extra practice or enrichment, including *Extra Practice for Primary Mathematics* (U.S. Edition), and *Primary Mathematics Intensive Practice* (U.S. Edition). These follow the same sequence as the *Primary Mathematics* (U.S. Edition) and can be used as a source of more problems or more challenge.

1 Numbers to 1000

Objectives

- ♦ Count, read, and write whole numbers to 1000 and identify the place value for each digit.
- ♦ Read and write number words for numbers to 1000.
- ♦ Count on or back by hundreds, tens, or ones from a number within 1000.
- ♦ Order and compare whole numbers to 1000 by using the symbols <, >, and =.
- ♦ Rename 10 ones as 1 ten and 10 tens as 1 hundred.
- ♦ Rename 1 ten as 10 ones and 1 hundred as 10 tens.
- ♦ Review addition and subtraction within 20.

Suggested number of weeks: 2-3

		TB: Textbook WB: Workbook	Objectives	Material	Appendix
1.1	**Looking Back**				
1.1a	Review: Tens and Ones	TB: pp. 6-8 WB: pp. 7-8	♦ Group objects by tens and ones in order to count them. ♦ Relate a 2-digit number to tens and ones. ♦ Read and write 2-digit numbers in numerals.	♦ Base-10 blocks ♦ Items that can be bundled into tens ♦ Place-value cards ♦ Place value charts	a14-a15
1.1b	Review: Number Words	TB: p. 8 WB: p. 9	♦ Read 2-digit numbers in words. ♦ Write 2-digit numbers in words. ♦ Rename tens and ones.	♦ Base-10 blocks ♦ Number cubes	
1.1c	Review: Count On and Count Back	TB: p. 9 WB: pp. 10-11	♦ Count within 100 by tens and ones. ♦ Count on 1, 2, 10, or 20 from a given number. ♦ Count back 1, 2, 10, or 20 from a given number.	♦ Hundred-chart ♦ Base-10 blocks ♦ Place-value cards	a14-a15 Mental Math 1
1.1d	Review: Addition and Subtraction within 20		♦ Review addition and subtraction facts within 10. ♦ Review mental math strategies for addition within 20. ♦ Review mental math strategies for subtraction within 20.	♦ Counters ♦ Dimes and pennies ♦ Number cards ♦ Math fact cards	Mental Math 2-8

		TB: Textbook WB: Workbook	Objectives	Material	Appendix
1.2	**Comparing Numbers**				
1.2a	Greater Than and Less Than	TB: pp. 10-11 WB: pp. 12-14	♦ Order and compare whole numbers to 100. ♦ Use the symbols <, >, and =.	♦ Base-10 blocks ♦ Number cards	a14-a15
1.2b	Practice	TB: p. 12	♦ Review.	♦ Number cards ♦ Number cards ♦ Number cubes	Mental Math 9
1.3	**Hundreds, Tens and Ones**				
1.3a	Hundreds Place	TB: pp. 13-16 WB: pp. 15-18	♦ Understand the hundreds place. ♦ Relate 3-digit numbers to hundreds, tens, and ones.	♦ Base-10 blocks ♦ Items in bundles of tens and ones from lesson 1.1a	a14-a15
1.3b	Place-Value Discs	TB: pp. 17-19 WB: pp. 19-20	♦ Relate 3-digit numbers to place-value discs. ♦ Rename hundreds, tens, and ones.	♦ 1 dime ♦ 10 pennies ♦ 1 ten-dollar bill ♦ 10 one-dollar bills ♦ Place-value charts and discs	
1.3c	Number Words for 3-digit Numbers	TB: pp. 13-19 WB: pp. 21-23	♦ Write 3-digit numbers in words. ♦ Interpret 3-digit numbers.	♦ Place-value charts and discs	
1.3d	Count On and Count Back	TB p. 20 WB p. 24	♦ Count on 1, 10, or 100. ♦ Count back 1, 10 or 100. ♦ Count on or back by ones, tens, or hundreds.	♦ Place-value charts and discs	Mental Math 10
1.3e	Compare and Order Numbers within 1000	TB p. 21	♦ Compare and order numbers within 1000. ♦ Practice.	♦ Place-value charts and disks ♦ Number cards ♦ Number cubes	

1.1 Looking Back

Objectives

- Count, read, and write whole numbers to 100 and identify the place value for each digit.
- Read and write number words for numbers within 100.
- Count on or back by tens or ones for a number within 100.
- Review addition and subtraction within 20.

Material

- Counters
- Base-10 blocks or multilink cubes
- Hundred chart
- 1 dime, 9 pennies
- Number cubes, 0-5 and 4-9, for each group
- Math fact cards, see lesson 1.1d
- Number cards, 4 sets of 1-10 for each group of students
- Straws and tape or paper clips or craft sticks and rubber bands
- Appendix pp. a14-a15 (place-value cards)
- Mental Math 1-8

Notes

In *Primary Mathematics* 1, students learned to relate 2-digit numbers to the place-value concept. This concept is reviewed in this chapter.

Students need to understand 2-digit numbers in terms of a part-whole model, tens and ones, and base-10 material or numerals on a place-value chart.

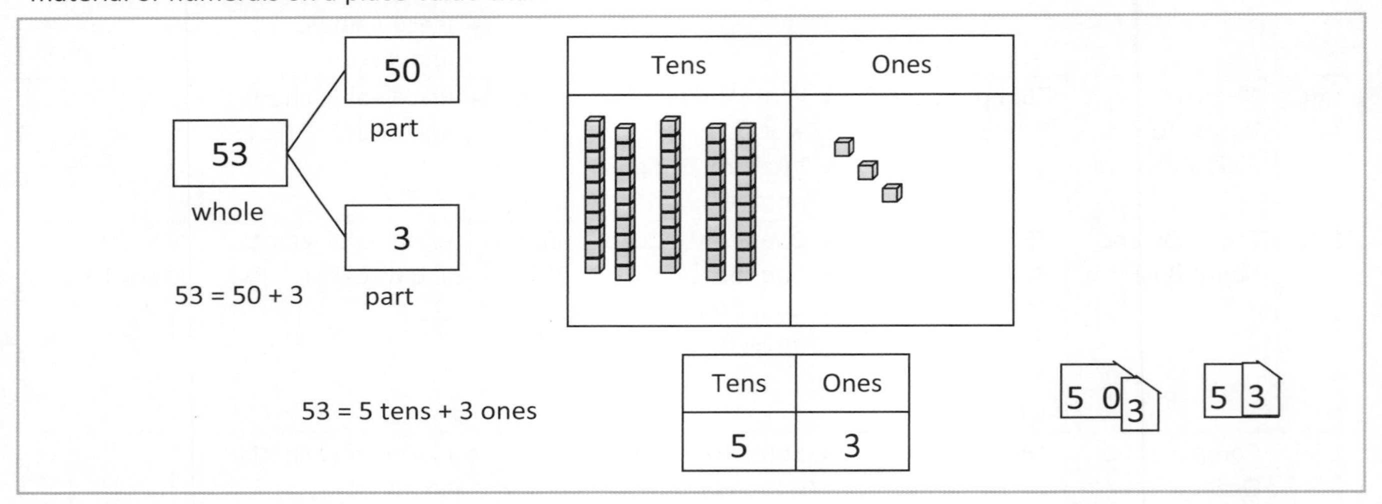

Place-value cards, such as those shown on p. 6 of the textbook and in the diagram above, can help students understand that the number representing the larger group (tens) comes first.

Number words are also reviewed in this part. However, do not let teaching how to spell the number words dominate the math lesson; it is not necessary for students to spell the number words correctly to proceed in math.

In *Primary Mathematics* 1A students learned to add and subtract within 20 through various strategies. They first learned number bonds for numbers within 10. For example, 3, 4, and 7 make a number bond; 3 and 4 are the parts and 7 is the whole. Then they committed the addition and subtraction facts through 10 to memory and used those facts to compute the addition and subtraction facts through 20. Lesson 1.1d in this guide is a review of addition and subtraction within 20, including the mental math strategies. If students have not learned these strategies, you should teach them as you proceed with this unit.

Even if students already know all the addition and subtraction facts within 20 by having simply memorized the math facts, understanding the strategy of making the next higher place or subtracting from the next higher place will be useful for mental math strategies in a variety of contexts and will increase their number sense and understanding of place value. Even students who are good with math concepts can have difficulty remembering math facts. The techniques for "making a ten" and "subtracting from a ten" will allow them to quickly calculate a math fact to 20 using the facts to 10, and know their answer is correct, rather than having to rely only on a possibly faulty memory.

The number bonds in the illustrations below are simply meant to illustrate the strategies. Drawing number bonds themselves should not become a new paper-and-pencil method for addition. Students can draw number bonds initially to aid them in their thinking process if they are weak on math facts through 10, but they should eventually and normally do these types of problems mentally.

⇒ Add the ones: To add ones to a 2-digit number when the sum of the ones is less than ten, add ones to ones.

```
  13 + 5 = 10 + 8 = 18
  / \
10   3
```

⇒ Make a 10: If the sum of the ones will be greater than 10, split one of the numbers up in order to "make a ten" with the other number.

```
8 + 5 = 10 + 3 = 13
   /\
  2  3
```

⇒ Subtract from the ones: To subtract ones from a ten and ones when there are enough ones, subtract the ones from the ones.

```
  18 – 5 = 10 + 3 = 13
  / \
10   8 – 5 = 3
```

⇒ Subtract from the 10: To subtract ones from a 2-digit number when there are not enough ones, subtract the ones from the tens and add that result to the ones.

```
 13 – 8 = 3 + 2 = 5
 / \
3   10 – 8 = 2
```

(A strategy not specifically taught but which students may come up with on their own and can use is to subtract the difference between the ones from 10. 8 – 3 = 5, 10 – 5 = 5. This method works because you are essentially first subtracting the 3 from 13, and then another 5; 3 and 5 is 8.)

⇒ Count on: When one of the numbers being added is 1, 2, or 3, count on from the other number to find the answer.

⇒ Count back: When the number being subtracted is 1, 2, or 3, count back from the other number to find the answer.

To practice mental math computation, you can have students do a "sprint" at the start of each lesson, using problems students need to practice. Other games and activities will be suggested in the appropriate lessons.

Sprint: Give students a sheet of paper with about 20 problems, such as those in the appendix. Have them solve as many as they can in 1 minute. Record the number they solved and the next day see if they can solve more problems of the same type in 1 minute. Students can see how much farther they get each day. If time permits, they can finish the rest of the problems on the sheet after the minute is over.

The mental math sheets in the appendix are a resource that can be used at any time after the lesson that refers to them.

1.1a Review: Tens and Ones

Objectives

- Group objects by tens and ones in order to count them.
- Relate a 2-digit number to tens and ones.
- Read and write 2-digit numbers in numerals.

Vocabulary

- Digits
- Place
- Tens place
- Ones place

Note

If some students have difficulty writing the numerals for 2-digit numbers correctly, let them use place-value cards such as those on appendix pp. a14-a15 to form the number first and then write the numerals. This will help them to write the tens first.

Task 2 on p. 7 includes the number 100, which requires a third place for 10 tens. You can point this out to students but do not dwell on it in this lesson. The hundreds place will be covered in Part 3.

Review: Place value within 20	
Write the numerals 1, 2, 3, 4, 5, 6, 7, 8, and 9 on the board. Ask students what these symbols represent. They tell us how many there are of something. Ask students what we use for "none" and then include the numeral 0. Tell them that we call these symbols *digits.* There are 10 digits.	1 2 3 4 5 6 7 8 9 0
Display 9 unit cubes or other objects or draw 9 circles on the board. Ask students to write the symbol for how many objects there are.	9
Now display 15 objects and ask students to write the number. After they write "15" ask them what each digit means. The 1 cannot mean 1, because there is more than 1 object. Circle 10 objects and explain that the 1 means there is 1 group of 10 objects. The 5 means that we have 5 ones. Explain that since we only have 9 digits for counting, making groups of tens gives us a way to use 9 digits for numbers greater than 9. Draw a place value chart with the number 15. Tell students that the *place* for the number of groups of tens is before the place for the number of ones. 1 is in the *tens place*, and 5 is in the *ones place*.	15 Tens | Ones 1 | 5
Remove or erase 5 of the objects and ask students to write the number of objects. Write 10 on the board in a place-value chart and then by itself. Tell students we have to write 0 in the ones place so that we know that the 1 is in the tens place and stands for a group of ten and not a one.	10 Tens | Ones 1 | 0
Discussion	**Text pp. 6-8**
There are 8 groups of tens and 3 ones. Write 83 on the board and discuss the meaning of each digit. Task 1: Have students write down the answers and say them aloud.	1. (a) There are **45** mangoes (b) 40 and 5 make **45**. (c) 5 more than 40 is **45**. (d) 40 + 5 = **45**.

Task 2: Have students count out loud by tens. Get them to count back by tens as well. Task 3: Have students write the answers. Emphasize that the number of groups of tens is written first, then the number of ones.	2. There are **100** stamps. 3. (a) 2 tens 6 ones = **26** (b) 4 tens 3 ones = **43** (c) **5** tens **7** ones = **57**
Assessment	
Display some numbers using ten-rods and unit-cubes and have students write the numbers. Do not always put the ten-rods to the left of the unit cubes. For more capable students, display some ten rods and more than 10 unit-cubes and ask students to write the numbers; they must form another ten from the unit-cubes. Have students do the following.	34
⇒ Write 64 as tens and ones.	6 tens 4 ones
⇒ Write 40 as tens and ones.	4 tens 0 ones
⇒ Write the number that is 4 more than 30.	34
⇒ 78 is 8 more than what number? Write it.	70
⇒ Write the number: 9 tens and 1 one.	91
⇒ Write the number: 3 ones and 5 tens.	53
Practice	WB Exercise 1, pp. 7-8, Problems 1-4

Activity

Divide students into groups. Give each group less than 100 straws and some tape, or craft sticks and rubber bands, or paper clips which they can string into tens. Ask them for ideas on how to count the objects without losing track. Lead them to suggest grouping them by tens. Then have them count by grouping the objects into tens. Assist them in bundling their groups of ten.

Have students record their results as tens and ones.

Repeat until each group has at least 9 tens bundled up. Save the groups of tens for the activity in lesson 1.3a.

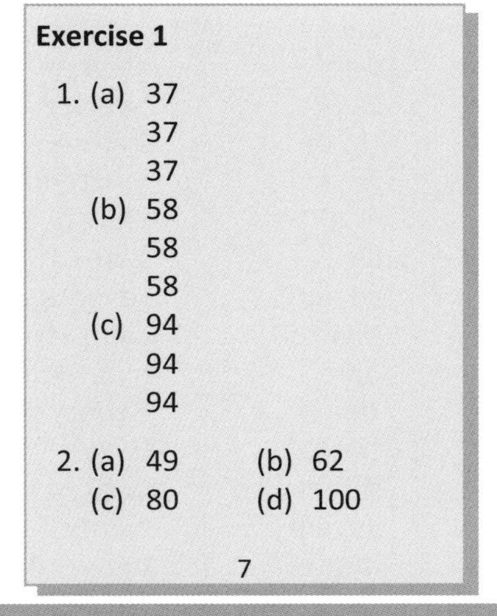

Exercise 1

1. (a) 37
   37
   37
   (b) 58
   58
   58
   (c) 94
   94
   94

2. (a) 49 (b) 62
   (c) 80 (d) 100

7

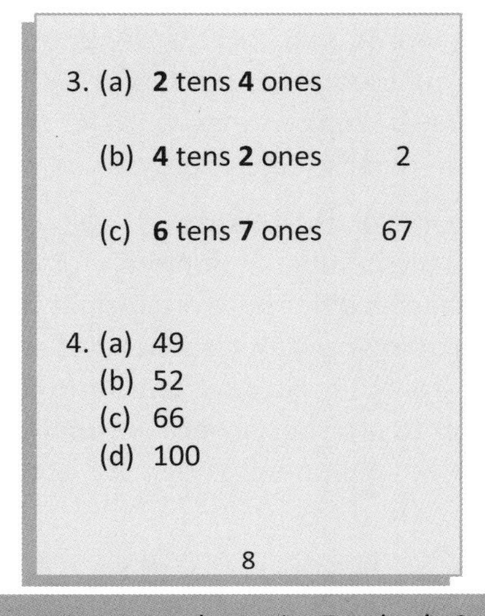

3. (a) **2** tens **4** ones
   (b) **4** tens **2** ones 2
   (c) **6** tens **7** ones 67

4. (a) 49
   (b) 52
   (c) 66
   (d) 100

8

1.1b Review: Number Words

Objectives

- Read 2-digit numbers in words.
- Write 2-digit numbers in words.
- Rename tens and ones.

Note

Some students will have trouble spelling number words. Rather than spending too much time during the math lesson on spelling and writing number words correctly, incorporate number words in spelling lessons. Other than in the lesson where it is taught, and in some of the review exercises, students will not have to be able to read and write number words to do the exercises.

Rename tens and ones	
Expand on Task 3(c) on p. 8. Draw the number bond. Remind students that a number bond shows two parts connected to a whole. In this number bond, one part is the tens, and the other part is the ones. Ask students for the corresponding addition equation. Have students show the number with base-10 blocks.	50 7 57 50 + 7 = 57
Give students the following problems. They can use base-10 blocks if needed. ⇒ 57 is 7 more than _____. ⇒ 50 is 7 less than _____. ⇒ 57 is 5 tens and _____ ones. ⇒ 57 is _____ ones. ⇒ 57 is 4 tens and _____ ones. ⇒ 57 is 3 tens and _____ ones. ⇒ 57 is ____ tens and 37 ones. ⇒ 57 is ____ tens and 47 ones.	 50 57 7 57 17 27 2 1
Review: Number words	
Refer to the number words on p. 8 in the textbook and have students read them out loud. Point out that there is a dash between the word for the tens and the word for the ones for numbers greater than twenty, e.g., twenty-six. Write the number words on the board and have students copy them and then write the number in numerals for each. Numbers between 10 and 20 have a different word format than tens and ones greater than 20. (You can tell students that even though numbers between 10 and 20 also have tens (one ten) and ones, the numbers through 18 are special because adding any two single digits together will give a number less than 20, and so they deserve special recognition.)	zero one eleven two twelve three thirteen four fourteen five fifteen six sixteen seven seventeen eight eighteen nine nineteen ten twenty thirty forty fifty sixty seventy eighty ninety hundred

Write a few number words for 2-digit numbers on the board and have students read them. Point out that with numbers past 20 we can write the number we "hear" first, such as 7 and then 1 for seventy-one. With teen numbers (13-19), though, we write the "teen" or ten first. Eleven and twelve are just different.	Seventy-one 71 Thirty-nine 39 Nineteen 19
Assessment	
Write the problems at the right and ask students to write the number and optionally the number words.	2 tens (20, twenty) 6 tens 3 ones (63, sixty-three) 4 tens 12 ones (52, fifty-two) 1 ten 80 ones (90, ninety) 4 ones 1 ten (14, fourteen)
Write some number words and have students write the numbers as tens and ones. Dictate some numbers for students to write both the number and number word.	Eleven (1 ten 1 one) Seventy-five (7 tens 5 ones)
Practice	WB Exercise 1, p. 9, Problems 5-6

Group Activity

Material: Number cubes 0-5 and 4-9 for each group.

Procedure: Students take turns throwing the dice and forming a 2-digit number which the other students write down, along with the corresponding number word.

Exercise 1

5. (a) 46
   (b) 67
   (c) 58
   (d) 93
   (e) 81
   (f) 25
6. (a) fifty
   (b) sixty-four
   (c) twenty-one
   (d) ninety-nine
   (e) thirty-two
   (f) one hundred

9

1.1c Review: Count On and Count Back

Objectives

- Count within 100 by tens and ones.
- Count on 1, 2, 10, or 20 from a given number.
- Count back 1, 2, 10, or 20 from a given number.

Note

If students have difficulties with the lesson, allow them to use base-10 blocks or place-value cards to form the numbers being discussed.

If students find the lesson easy, spend more time having them count on and count back when the tens change, such as 71 – 2 or 49 + 2. Also include counting on or counting back by 3 ones and 3 tens. Use Mental Math 1.

Be sure students do not include the given number when adding or subtracting by counting on.

<table>
<tr><td>Review number order</td><td></td></tr>
<tr><td>Display a hundred-chart. Let students follow along with individual charts.

Discuss the order of the numbers on the chart. The numbers increase from left to right. The number after the last one in the row is at the start of the next row.

Ask students to point out any patterns they see. For example, all the numbers in each column have the same digit in the ones place, while the digit in the tens place increases. The numbers along a row have the same digit in the tens place, except for the last one. Each row is a group of ten.

Point to the number 53, and ask students which number is one less, one more, ten less, or ten more than 53.

Then repeat with 60 and 81.</td><td>
<table>
<tr><td>1</td><td>2</td><td>3</td><td>4</td><td>5</td><td>6</td><td>7</td><td>8</td><td>9</td><td>10</td></tr>
<tr><td>11</td><td>12</td><td>13</td><td>14</td><td>15</td><td>16</td><td>17</td><td>18</td><td>19</td><td>20</td></tr>
<tr><td>21</td><td>22</td><td>23</td><td>24</td><td>25</td><td>26</td><td>27</td><td>28</td><td>29</td><td>30</td></tr>
<tr><td>31</td><td>32</td><td>33</td><td>34</td><td>35</td><td>36</td><td>37</td><td>38</td><td>39</td><td>40</td></tr>
<tr><td>41</td><td>42</td><td>43</td><td>44</td><td>45</td><td>46</td><td>47</td><td>48</td><td>49</td><td>50</td></tr>
<tr><td>51</td><td>52</td><td>53</td><td>54</td><td>55</td><td>56</td><td>57</td><td>58</td><td>59</td><td>60</td></tr>
<tr><td>61</td><td>62</td><td>63</td><td>64</td><td>65</td><td>66</td><td>67</td><td>68</td><td>69</td><td>70</td></tr>
<tr><td>71</td><td>72</td><td>73</td><td>74</td><td>75</td><td>76</td><td>77</td><td>78</td><td>79</td><td>80</td></tr>
<tr><td>81</td><td>82</td><td>83</td><td>84</td><td>85</td><td>86</td><td>87</td><td>88</td><td>89</td><td>90</td></tr>
<tr><td>91</td><td>92</td><td>93</td><td>94</td><td>95</td><td>96</td><td>97</td><td>98</td><td>99</td><td>100</td></tr>
</table>

53 1 less: 52 1 more: 54
10 less: 43 10 more: 63

60 1 less: 59 1 more: 61
10 less: 50 10 more: 70

81 1 less: 80 1 more: 82
10 less: 71 10 more: 91</td></tr>
<tr><td>Count by tens and ones</td><td></td></tr>
<tr><td>Use the hundred-chart. Point to a number that is not a multiple of ten, such as 16, and have students count on from that number by tens.</td><td>16, 26, 36, 46, 56, 66, 76, 86, 96</td></tr>
<tr><td>Point to another number, such as 88, and have students count back from that number by tens.</td><td>88, 78, 68, 58, 48, 38, 28, 18, 8</td></tr>
<tr><td>Circle two numbers, such as 31 and 75, and ask students to decide which number is greater based on their positions on the chart.

Then have them count on or back from one number to the next, first by tens, then by ones, as you point to the numbers on the chart.

Have students count on or back between two numbers, first by tens and then by ones, without the chart.</td><td>Forward: 31, 41, 51, 61, 71, 72, 73, 74, 75.

Backward: 75, 65, 55, 45, 35, 34, 33, 32, 31.</td></tr>
</table>

Discussion	Text p. 9	
Tasks 4-5: Show the number 65 with base-10 blocks. As students supply the answers, ask them whether we should add or remove unit-cubes or ten-rods.	4. (a) 66 (b) 64 (c) 75 (d) 55	
Task 6: Have students do these without a hundred chart or base-10 blocks.	5. (a) 67 (b) 63 (c) 85 (d) 45	
	6. (a) 81 (b) 82 (c) 90 (d) 100 (e) 79 (f) 78 (g) 70 (h) 60	
Assessment		
Write some addition and subtraction expressions that can be solved by counting on or back by 1, 2, 10, or 20 and ask students for the answers.	74 + 1	(75)
	67 – 1	(66)
	33 + 10	(43)
	49 + 2	(51)
	81 – 2	(79)
	56 – 20	(36)
Enrichment	Mental Math 1	
Practice	WB Exercise 2, pp. 10-11	

Activity

Face the students and hold your hands out face up. Tell them that your right hand (which will be on their left) is tens, and your left hand is ones. Tell them that when you push up with the tens hand they should count up by ten, and when you push down with it they should count down by ten. When you push up with the ones hand they should count up by one, and when you push down with it they should count down by one. Start at a random number and count with them as you move your hands. Be careful not to mix hands if you are facing them, your right is their left and tens should be on their left.

Exercise 2

1. (a) 77
   (b) 75
   (c) 86
   (d) 66
2. (a) 78
   (b) 74
   (c) 96
   (d) 56
3. (a) 40
   (b) 73
   (c) 100
   (d) 73
   (e) 76
   (f) 74

10

4. (a) 56 (b) 57
   (c) 65 (d) 75
   (e) 54 (f) 53
   (g) 45 (h) 35
5. (a) 71 (b) 72
   (c) 80 (d) 90
   (e) 69 (f) 68
   (g) 60 (h) 50
6. (a) 49 (b) 50
   (c) 58 (d) 68
   (e) 47 (f) 46
   (g) 38 (h) 28

11

1.1d Review: Addition and Subtraction within 20

Objectives

- Review addition and subtraction facts within 10.
- Review mental math strategies for addition within 20.
- Review mental math strategies for subtraction within 20.

Note

Students should already know addition and subtraction facts through 10, and probably through 20. However, the strategies shown in this lesson, which were taught in *Primary Mathematics* 1A, will help them if they forget a fact and are applicable to strategies which will be taught later. They enhance a student's number sense and understanding of place-value. Some students are very good at math concepts, but do not memorize math facts well. These strategies will allow them to quickly add and subtract within 20 by using facts within 10.

Review: Facts to 10	Mental Math 2-4
Make sure students know the addition and subtraction facts to ten. You can quiz them orally or use Mental Math 2-4 in the appendix.	
Add by making a ten	
Write the expression **7 + 5** on the board. Draw a 2 by 5 grid and draw circles or place counters in 7 spaces (filling up 5 in a row first). Draw or show 5 more counters outside of the grid. Ask students how many more are needed to make a ten. Three more are needed, so move three counters into the empty spaces. Ask students how many counters are left over as ones. There are two. So 7 + 5 is the same as 10 and 2. Show students the working with a number bond, splitting 5 into 3 and 2.	7 + 5 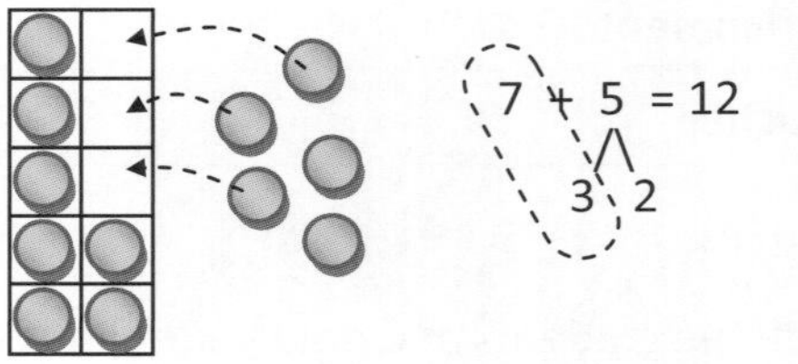 7 + 5 = 12 3 2
Write the expression **4 + 9** on the board. Ask for suggestions on how to make a ten. Point out that we can either make a ten with the 4 or with the 9 and show both ways with number bonds. Usually it is easier to make a ten with the larger number.	4 + 9  4 + 9 = 13 3 1 4 + 9 = 13 6 3
Repeat with other examples as needed, allowing students to use a grid and counters if needed. Write some addition problems and ask how many are needed to make a ten with one of the digits, and how many are left over.	8 + 5 = 10 + ____ (3) 7 + 4 = 10 + ____ (1) 6 + 9 = 10 + ____ (5)
Subtraction strategies	
Write the expression **18 – 5**. Illustrate the number 18 with ten counters on a grid and 8 off the grid. Ask students how we can subtract 5. We can simply subtract 5 ones from 8 ones. 8 – 5 is 3, so 18 – 5 is the same as 10 and 3. The answer includes the ten that we still have. Show the process with a number bond, splitting 18 into 10 and 8.	18 – 5 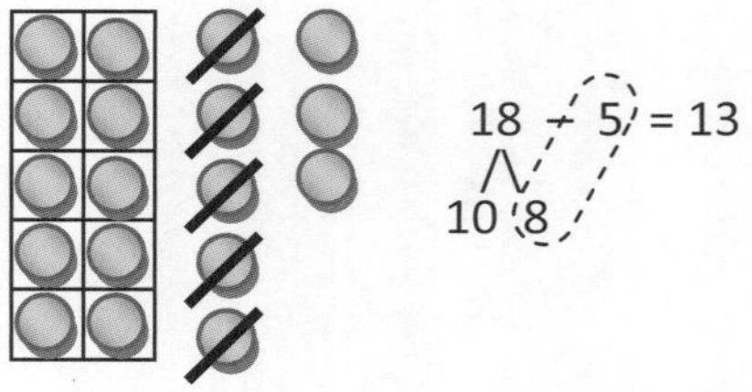 18 – 5 = 13 10 8

Write the expression **10 – 8** and have students supply the answer.	10 – 8 = 2
Now write the expression **14 – 8**. Illustrate 14 with the grid and counters. Ask for suggestions on how to subtract 8. Point out that that we cannot take 8 from the 4 ones because there are not enough ones, but we can take it from the ten. Remove 8 counters and ask how many are left. There are 2 from subtracting 8 from 10, and 4 ones that we started with, so there are 2 + 4 = 6 counters left. So 14 – 8 is 6. Show the process with a number bond as well, splitting the 14 into 4 and 10. Ask for alternative strategies.	14 – 8 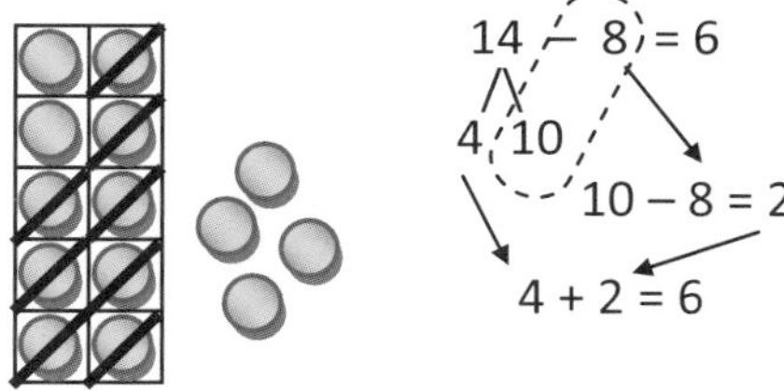
We can also mentally subtract in 2 steps. We can first take away the 4 ones. Ask how many more we still need to take away. Since 8 – 4 = 4, we need to take away an additional 4, leaving 6. So we can find the answer by first subtracting 4 from 8, and then subtracting that answer from 10. Repeat with other examples as needed.	14 – 8 = 14 – 4 – 4 (8 = 4 + 4) = 10 – 4 = 6
Reinforcement	
Show students a dime and 3 pennies. Ask how much money you have and write down 13 cents. Then tell students that you want to buy something that costs 7 cents. How would you pay for it? Show that you would use the dime, get 3 pennies change, so the total money you have left is 3 + 3 = 6 pennies.	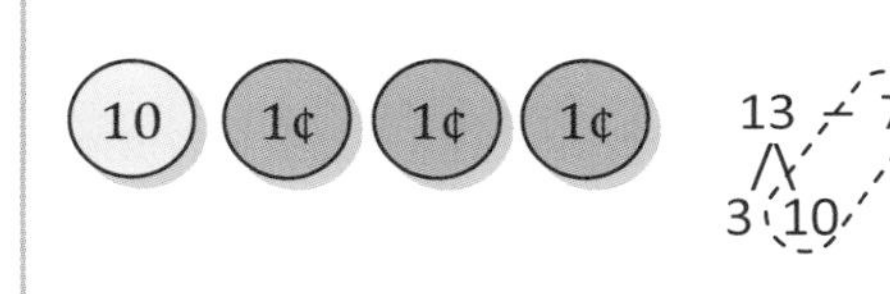
Mental math	Mental Math 5-8

Class game

Material: Cards with facts being practiced that the teacher can hold up in front of the class. Cards with answers, enough so that each student gets 5 cards.

Procedure: Students place their cards on their desks face up. The teacher holds up the fact cards one at a time. In this case, you could use addition and subtraction facts within 20 where renaming occurs. If a student has a card containing the answer to the fact card, he or she turns the card over (to face down). The student that turns over all five cards first wins and can have a turn holding up the fact cards.

Group game

Material: Number cards 1-10, 4 sets for each group of students.

Procedure: Deal out all cards face down.

Game 1: Each player turns over 2 cards and adds the numbers. The player with the greatest total gets all the cards that have been turned over. If the total is the same then the player with the highest card gets the cards. The player with the most cards after all cards have been turned over wins.

Game 2: Each player turns over 2 cards and adds 10 to the number on the first card to get a number between 10 and 20. The player then subtracts the value on the second card from this number. The student with the lowest answer gets all the cards that have been turned over. If the answer is the same for both players, the player with the card with the lowest number gets the cards. The player with the most cards after all cards have been turned over wins.

1.2 Comparing Numbers

Objectives

- Order and compare whole numbers to 100.
- Use the symbols <, >, and =.

Material

- Base-10 blocks for board and for students
- Large number cards 1-99, and symbol >
- Number cards 0-9, 4 sets per group
- Number cubes, 1 per group
- Mental Math 9
- Number cards 1-99

Prerequisites

Students need to understand the concept of place-value for tens and ones.

Notes

In *Primary Mathematics* 1B and in the previous lessons, students learned to compare and order numbers based on their position in a hundred-chart. Here, they will compare numbers by focusing on the tens and ones digits of the numbers they are comparing. The symbols for greater than (>) and less than (<), will be introduced.

Numbers can be compared by comparing the digits in the highest place value first. If they are the same, then we compare the digits in the next highest place value, and so on. This process will be extended in the next part to numbers within 1000 and in later levels to numbers past 1000.

For numbers within 100, we first compare the tens. The number with the greater digit in the tens place is the greater number. If the digits in the tens place are the same, then we compare the ones. The number with the greater digit in the ones place (but the same digit in the tens place) is the greater number.

Students will first compare numbers using base-10 material, and then by simply looking at the numbers. They need to realize that they are comparing tens, not just the first digit in the number. With numbers that have only 1 or 2 digits, this is fairly obvious from being familiar with a hundred-chart and the order of numbers on it, but with numbers with more digits, such as students will encounter in later grades, it may be less obvious.

Compare 21 and 12.	Compare 32 and 35.	Compare 25 and 8.
2 1 1 2 2 tens > 1 ten, so 21 > 12	3 2 3 5 3 tens = 3 tens 2 ones < 5 ones, so 32 < 35	2 5 8 2 tens > 0 tens, so 25 > 8

Listed below are some additional games or suggestions for ongoing math fact practice. Some can be adapted for multiplication facts later.

- Use large fact cards and cards with the answers on a loop of string or masking tape folded over on the back so that the students can display their cards but leave their hands free. There should be as many fact cards and answer cards as students. Each student gets an answer card. They line up in a horseshoe shape in the numerical order of their answer cards. Stand in the opening of the horseshoe and show a fact card. Say the answer out loud and point to the student wearing the correct answer. Show another fact card. The student first pointed to must say the answer out loud and point to the student wearing the correct answer to the new fact, who is now the pointer for the next fact card. Students see how fast they can keep the game going. Once they get used to the game, they can do it silently, just pointing to the student with the answer.

- Use several sets of number cards 1-10 large enough to be seen by the whole class. Divide students into about 4 teams and line each team up. The first student in each line comes to the front of the room and draws two cards and adds the numbers on them. The student with the greatest (or lowest) answer collects all the students up front to now be part of his or her team. They go back to the end of the line for their team. Play continues for a while; the team with the most members at the end wins.

- Divide students into groups. Provide each group with four sets of number cards 1-10. One student turns over two cards. Other students must add the numbers together. The student who gets the correct answer first turns over the next two cards.

- Divide students into groups. Provide each group with four sets of number cards 1-9 and 2 number cubes labeled with 4-9. Each group chooses a dealer who throws the number cubes and adds the numbers that end up on the cubes' tops. This sum is the target number. The dealer then draws one card at a time and shows it to the other students. The students must subtract the number on the card from the target number. For example, the dealer gets 15 as the target number and draws an 8. The correct answer is 7. The student who gets the correct answer first becomes the next dealer or students can take turns being the dealer.

1.2a Greater Than and Less Than

Objectives

- Order and compare whole numbers to 100.
- Use the symbols <, >, and =.

Vocabulary

- Symbol
- Is greater than
- Is less than

Note

If some students have difficulty comparing 2-digit numbers without use of manipulatives to visually see when the tens are greater, have them rewrite the numbers vertically, one above the other on a place-value chart with columns for tens and ones. Students can also use place-value cards.

For problem 6 in workbook Exercise 3, allow students to write the symbol rather than the words in the blank.

See lesson 1.2a for activities that could be used with this lesson.

Introduce the symbols > and <	
Write **2 + 3 = 5** and below that **5 = 5** but without the equal signs. Ask students what symbol we would use to show that 2 + 3 is the same as 5. Insert the equal symbol. Point to the "+" and the "=" and say that we can call these *symbols*. A symbol stands for something else. The symbol "+" stands for *plus* or *added to*. The symbol "=" stands for *is equal to*.	2 + 3 5 5 5 2 + 3 = 5 5 = 5
Now write the numerals **5** and **3** and ask students whether these two are equal. They are not. Ask them whether 5 is greater than or less than 3. Then write the symbol ">" between the two numbers. Tell students that this symbol stands for *is greater than*.	5 3 5 > 3
Write the numerals **1** and **6** on the board. Ask whether 1 is greater than or less than 6. Write the symbol "<" between the two numbers and tell students that this symbol stands for *is less than*.	1 6 1 < 6
Discussion	**Text p. 10**
Discuss page 10 in the text. Tell students that they can think of the mouth of a greedy crocodile that eats the greater number to help them decide which way the symbols for greater than and less than should point. Point out that 21 is greater than 12 because it has more tens. 99 is less that 100 because 99 has 9 tens but 100 has 10 tens.	
Compare numbers	
Draw or display base-10 blocks to represent the numbers as needed in the following discussion. You can put them on a place-value chart, one number above the other. You can provide students with base-10 blocks to follow along with the discussion and show the numbers rather than or as well as showing them on the board. Write two 2-digit numbers with different tens, one above the other, with the digits aligned. Have students show both numbers with base-10 material. Point out that the number with more tens is larger. Write the numbers next to each other with the symbols between them.	Tens | Ones 32 52 32 < 52 52 > 32

Repeat with two numbers with the same tens digit but different ones digits, such as 56 and 53. Tell students that we first compare the tens. If the tens are the same, we then compare the ones to see which number is greater.	5 6 56 > 53 5 3 53 < 56
Repeat with a 1-digit number and a 2-digit number. Point out that the first digit of the 1-digit number is greater, but it is ones. The 9 has no tens. 12 is greater because it has more tens.	9 9 < 12 1 2 12 > 9
Write some numbers next to each other and have students tell supply the symbols for *greater than*, *less than*, or *equal to* between them.	32 < 59 82 = 82 60 > 38 9 < 17 46 < 48 100 > 10
Assessment	**Text p. 11**
Task 3: Students should first determine that two of the numbers are greater than the other two according to the tens position, and then put each pair in order according to the ones digit.	1. (a) 43 > 34 (b) 69 < 78 (c) 35 > 32 (d) 29 < 37 (e) 47 < 50 (f) 50 > 49 2. (a) 39 (b) 30 (c) 56 (d) 98 3. 50, 59, 90, 95
Reinforcement	
Use number cards 1-99. Students can work in groups. Give each group around 10 random cards to put in order.	Number cards 1-99
You can prepare large cards with numbers between 1 and 99, enough for each student have 1 card (so not all 99 numbers are used). Get students to line up in order according to the numbers on their cards.	Large number cards 1-99
Practice	WB Exercise 3, pp. 12-14

Exercise 3

1. (a) 50
   (b) 59
   (c) 28   (d) 70
   (e) 87   (f) 100

2. (a) 45   (b) 87
   (c) 63   (d) 100
   (e) 70   (f) 57

12

3. (a) 23   (b) 24
   (c) 29   (d) 78
   (e) 54   (f) 87
   (g) 60   (h) 98

4. (a) 31   (b) 50
   (c) 45   (d) 56
   (e) 15   (f) 36

13

5. (a) 67, 76, 78, 87
   (b) 90, 82, 79, 66

6. (a) greater than [>]
   (b) less than [<]
   (c) greater than [>]
   (d) less than [<]
   (e) less than [<]
   (f) greater than [>]
   (g) greater than [>]
   (h) less than [<]
   (i) less than [<]
   (j) greater than [>]

14

1.2b Practice

Objectives

- Review.

Note

In this guide practices will be considered a lesson. They do not have to be used that way. If you are behind schedule, they could be used as homework, provide additional material for a lesson, or they can be done as you proceed with the lessons for more continuous review. In some cases, the problems in the practices go into the concepts more deeply, and can be used to provide additional teaching, particularly for word problems, so even if you use them as homework it is important to go over at least some of the problems with the students.

Practice	**Text p. 12, Practice 1A**
This practice can either be done by the class as a whole, or students can work on the problems individually as you circulate and provide assistance. Problem 7(e): Students should be able to solve this one by extending the knowledge they already have, even though one of the numbers is greater than 100. They have seen the number 100, and they know that the two numbers are the same except for the ones place, so all they have to compare is the number in the ones place. If students have trouble, read the numbers aloud and point out that you need to count on from 100 to get to 105, so 105 is greater, or simply save this problem for later.	1. (a) forty-four (b) fifty-five (c) ninety-five (d) one hundred 2. (a) 6 tens 5 ones (b) 4 tens 0 ones (c) 7 tens 8 ones (d) 9 tens 7 ones 3. (a) 66 (b) 81 (c) 53 (d) 70 4. (a) 54 (b) 73 5. (a) 100 (b) 49 6. (a) 89 (b) 35 7. (a) $34 > 29$ (b) $89 < 90$ (c) $46 > 45$ (d) $71 > 70$ (e) $105 > 100$ (f) $50 < 52$
Mental math practice	Mental Math 9

Activity

Material: Large number cards 1-99, symbol card > (which, upside down, is <).

Procedure: Call three students up front. Give two of them each a number card. Give the third student the symbol card (crocodile card). The third student is the crocodile and must stand between the two students and orient the card correctly so that the open part of the symbol faces the student with the greater number. That student is "eaten" and must sit down. A new student comes up and gets the symbol card, while the student who originally had the symbol card gets a number card. Play continues until all students have been the crocodile.

Group game

Material: Number cards 0-9, 4 sets per group.

Procedure: Each player draws two cards. The first card is the tens, the second the ones. They then compare their numbers. The player with the greater number gets all the cards. If two players tie, they must draw again. The player with the most cards at the end wins.

Group game

Material: Number cube 1-6 or 10-sided cube for each group.

Procedure: Students draw two lines or a place-value chart for tens and ones on their paper. Each student rolls the number cube once, decides whether the number thrown should be a ten or a one, and writes it in the appropriate place. Then each student rolls the number cube a second time and writes the number in the remaining place. The student with the highest 2-digit number at the end of a round gets a point. Students can also record all of their group's numbers in order for each round.

1.3 Hundreds, Tens and Ones

Objectives

- Count, read, and write whole numbers to 1000 and identify the place value for each digit.
- Represent numbers within 1000 with place-value discs on a place-value chart.
- Count on or back by hundreds, tens, or ones for a number within 1000.
- Rename 10 tens as 1 hundred or 1 hundred as 10 tens.
- Compare and order numbers within 1000.

Material

- Base-10 blocks
- Items in bundles of tens and ones from lesson 1.1a
- 1 dime, 10 pennies, 1 ten-dollar bill, and 10 one-dollar bills
- Place-value charts and discs
- Place-value cards
- Number cards
- Number cubes
- Mental Math 10

Prerequisites

Students should thoroughly understand place-value concepts for numbers within 100, as covered in the previous lessons.

Notes

In this part students will extend their understanding of place value to hundreds.

Up until now students have been using base-10 material where they can see that a ten is composed of ten ones, such as the ten-rod of base-10 blocks. In this part students will first relate place value to dollar bills to see that a ten-dollar bill represents the same amount of money as ten one-dollar bills. Then they will start using place-value discs to represent numbers.

Place-value discs are used in *Primary Mathematics* to illustrate concepts involving the base-10 number system. These are round discs with 1, 10, or 100 written on them. Students will learn that 1 ten-disc represents the same value as 10 one-discs, and 1 hundred-disc represents the same value as 10 ten-discs or 100 one-discs. They will be able to illustrate all numbers within 1000 with the number discs.

With place-value discs, the number on the disc or its placement on a place-value chart indicates its value; students do not see 10 one-discs within a ten-disc. It is thus a more abstract way to represent numbers than using base-10 blocks. The discs are easier to work with than base-10 blocks and in later levels will be extended to represent numbers greater than 1000 as well as decimal numbers. After using concrete discs, students can simply draw circles in the appropriate column on a place-value chart.

Students will learn to count on or back by 1, 10, or 100. They will use place-value charts and place-value discs to see what happens when counting on or back 1 or 10 requires renaming. For example, 10 more than 394 is 404. If we show 394 on the chart with discs in the corresponding columns, then in order to add one more ten we need to replace, or rename, 10 tens as 1 hundred in order to write the number.

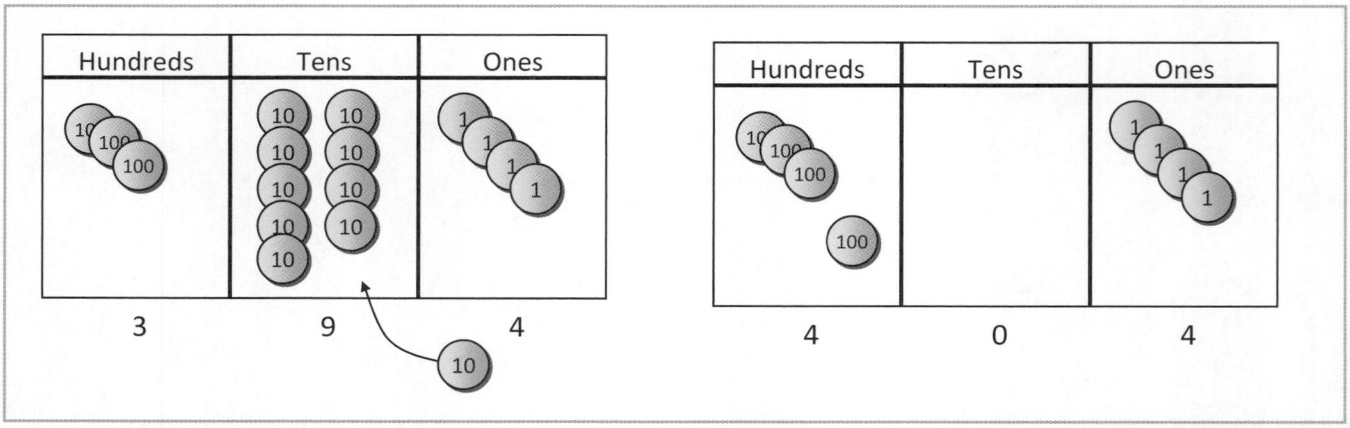

Hundreds	Tens	Ones
100 100 100	10 10 10 10 10 10 10 10 10	1 1 1 1
3	9	4

(+ 10)

Hundreds	Tens	Ones
100 100 100 100		1 1 1 1
4	0	4

Similarly, 10 less than 502 is 492. If we show 502 on the chart with 5 hundred-discs in the hundreds column, then in order to show 10 less, we need to rename 1 hundred as 10 tens before we can remove a ten. This process will help prepare students for adding and subtracting 3-digit numbers with renaming in the next unit.

Hundreds	Tens	Ones
100 100 100 100 100		1 1
5	0	2

Hundreds	Tens	Ones
100 100 100 100	10 10 10 10 10 10 10 10 10	1 1
4	9	2

10

Students will also compare 3-digit numbers, first by looking at the numbers with place-value discs, and then by looking at the digits in each place. To compare numbers we first compare the digits with the highest place value. If they are the same we then compare the digits with the next highest place value, and so on.

5 4 0 4 6 5	3 8 0 3 4 1	7 3 2 4 1
5 hundreds > 4 hundreds, so 540 > 465 465 < 540	3 hundreds = 3 hundreds 8 tens > 4 tens, so 380 > 341 341 < 380	2 hundreds > 0 hundreds, so 241 > 73 73 < 241

1.3a Hundreds Place

Objectives

- Understand the hundreds place.
- Relate 3-digit numbers to hundreds, tens, and ones.

Vocabulary

- Hundreds
- Hundreds place
- One thousand

Note

If students have difficulty writing 3-digit numbers correctly or understanding them, let them use place-value cards to form the number first. This will help them visualize how the written number relates to the value in each place.

Introduce the hundreds place	
Draw a place-value chart with two columns for tens and ones on the board. Place or draw ten-rods and unit-cubes on the chart to show 99 and ask students what number is represented. Write the numbers below the chart. Then put one more unit-cube in the ones column and ask how many there are. Lead students to see that since there is no digit we can use to write 10 ones we need to bundle up the 10 ones into a ten. But if we put the ten in the tens column, there is also no digit to write that there are 10 tens. So we need another place, the hundreds place. Add a column to the chart, label it "Hundreds," replace the 10 tens with a hundred-flat, and put that in the hundreds column. Write the number below it. Tell students that a digit in the hundreds place tells us how many groups of hundreds we have. 1 hundred is a group of 10 tens. 100 means one hundred, zero tens, and zero ones. It is the way we can write a number to show one hundred without needing more numerals besides the ten digits.	Tens	Ones 9	9 Hundreds	Tens	Ones 1	0	0
Ask students how many tens and how many ones are in one hundred.	1 hundred = 10 tens 1 hundred = 100 ones
Discussion	**Text pp. 13-14**
Page 13: The boy is bundling straws first into tens, and then bundling 10 tens into hundreds. Have students count the hundreds out loud. Page 14: Guide students in counting the number of straws out loud, first by hundreds, then by tens, then by ones, as the children on the page are doing. For each number, ask how many hundreds, tens, and ones there are.	106 = 1 hundreds 0 tens 6 ones 140 = 1 hundreds 4 tens 0 ones 223 = 2 hundreds 2 tens 3 ones

Top of page 15: Tell students that for 10 hundreds, we need another place, which is the thousands place. A 1 in the thousands place means we have 10 groups of hundreds.	
Ask students how many groups of hundreds, tens, and ones are in 1000.	1 thousand = 10 hundreds 1 thousand = 100 tens 1 thousand = 1000 ones
Assessment	**Text pp. 15-16**
	1. (a) 3 hundreds 4 tens 6 ones = **346** (b) **4** hundreds **3** tens **7** ones = **437** 2. 236 3. (a) 258 (b) 470 (c) 809
Practice	WB Exercise 4, pp. 15-18

Activity

<u>Material</u>: Bundles of tens and ones from the activity in lesson 1.1a.

<u>Procedure</u>: Divide students into groups and give each group some of the bundles of tens along with more singles. Have them count how many they have by making more tens, and combining tens into hundreds. Try to divide groups and straws such that the total number at the end is less than 1000. When each group has found how many they have, have the class find out how many total straws there are among all the groups.

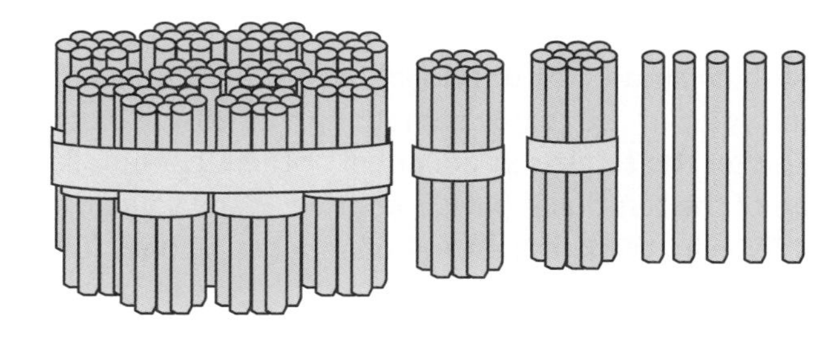

Exercise 4

1. (a) 214
   (b) 346
   (c) 305
   (d) 472
   (e) 563
   (f) 660
   (g) 790
   (h) 307

15

2. 355
   129
   219
   553
   704
   535
   740

16

3. (a) 175
   (b) 253
   (c) 240
   (d) 407

17

4. 611 309
   293 390
   90 6
   500 60

18

Place-Value Discs

Objectives

- Relate 3-digit numbers to place-value discs.
- Rename hundreds, tens, and ones.

Vocabulary

- Value

Note

New coins and bills have come into circulation, so some coins and bills in a classroom imitation set may not have the exact same appearance as those in the text, and/or those in the text may not have the same appearance as some real coins and bills. Students should still be able to extrapolate from general appearance.

Relate place value to coins and bills

Show students 1 dime and 10 pennies. Tell them that you want to buy something that costs 10 cents. Which coins can you use? You can either pay with the ten pennies, or with the dime. Explain that the dime has the same *value* as 10 pennies.

Now show students 1 ten-dollar bill and 10 one-dollar bills. Ask which you can use to buy something that costs 10 dollars. You can use either the single ten-dollar bill or the 10 one-dollar bills. The value of the ten-dollar bill is the same as the value of 10 one-dollar bills. We can think of the ten-dollar bill as 10 one-dollar bills bundled up together.

If you have a hundred-dollar bill available from a classroom money set, show it to students and ask what you could use instead of a hundred-dollar bill. To buy something that costs 100 dollars, we could use a hundred-dollar bill, or 10 ten-dollar bills. Students may suggest other combinations, such as 5 twenty-dollar bills.

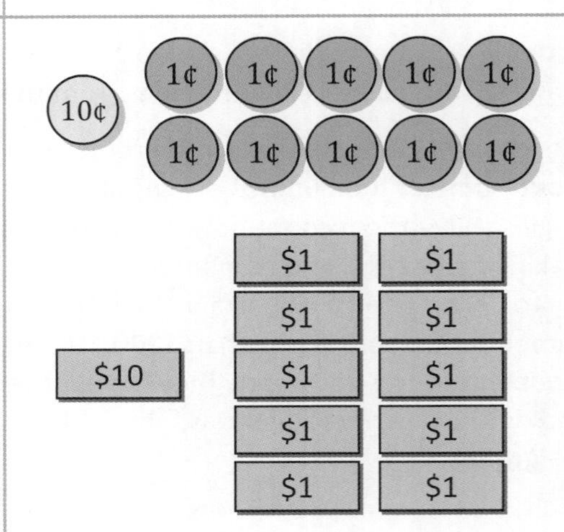

Discussion

Task 4: After students complete this task, discuss other ways we could buy something for $460, using only one-dollar, ten-dollar, and hundred-dollar bills. For example: If we have only 3 hundred-dollar bills, how many ten-dollar bills would we need? Even though we might have 3 hundred-dollar bills and 16 ten-dollar bills, how would we write how much money we have? We would still write the amount as $460. We could also use 10 one-dollar bills in place of a ten-dollar bill, or 20 one-dollar bills in place of two of the ten-dollar bills. In each case, the value of the money is the same, and how we write the value in numbers is the same.

Task 5: Ask for other combinations that will have the same value as a thousand-dollar bill, again assuming that we only have hundred-dollar, ten-dollar, and one-dollar bills. (We do not have any twenty-dollar or five-dollar bills.)

Text pp. 17-18

4. (c) 10
5. 10

$460 = 4 hundred-dollar bills
6 ten-dollar bills
0 one-dollar bills

$460 = 3 hundred-dollar bills
16 ten-dollar bills
0 one-dollar bills

$460 = 4 hundred-dollar bills
4 ten-dollar bills
20 one-dollar bills

Task 6: Tell students that instead of pictures of bills, we can simply draw circles or discs, and write the value in them. A disc marked with a 10 has the same value as 10 discs marked with a 1. We can imagine the ten-disc is made up of 10 one-discs bundled up together. Similarly, a hundred-disc is the same as 10 ten-discs, and a thousand-disc is the same as 10 hundred-discs.	6. (a) **100** ones = 1 hundred (b) **1000** ones = 1 thousand
Place-value discs	
Provide students with a place-value chart and place-value discs. Write some numbers on the board, such as those at the right, and have students show the number on the chart.	326 284 406 460 46
Assessment	**Text p. 19**
	7. **6** hundreds **2** tens **3** ones 8. (a) 467 (b) 250 (c) 306
Enrichment	
Ask students the following questions. Allow them to work out the answers with place-value discs on a place-value chart, if needed.	
⇒ How many 1's are in 10 tens?	100
⇒ How many 1's are in 10 hundreds?	1000
⇒ How many 1's are in 30 tens?	300
⇒ How many 1's are in 32 tens	320
⇒ How many 10's are in 400 ones?	40
⇒ How many 10's are in 460 ones?	46
⇒ How many 100's are in 50 tens?	5
Practice	WB Exercise 5, pp. 19-20

Exercise 5

1. (a) $460

(b) $303

(c) $339

19

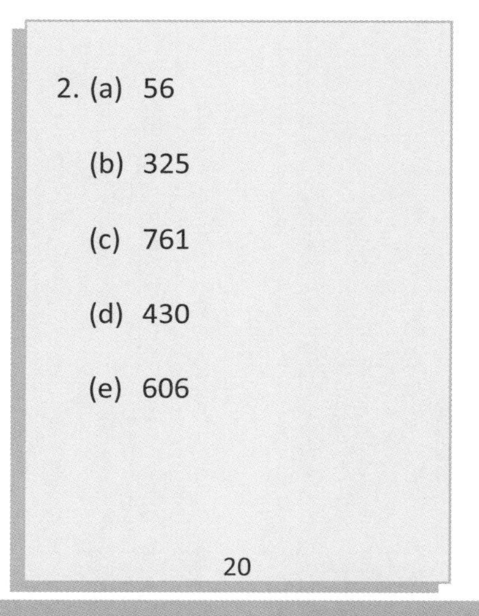
2. (a) 56

(b) 325

(c) 761

(d) 430

(e) 606

20

1.3c Number Words for 3-digit Numbers

Objectives

- Write 3-digit numbers in words.
- Interpret 3-digit numbers.

Note

Other than in Exercise 6 and in some of the review, students will not have to read and write number words. So for students who have difficulty with spelling the words, you can include the words in spelling lessons to provide additional practice at a different time than during math class.

Read and write number words	**Text pp. 13-17**
Draw students' attention to the number words on pages 13-17 in the text and have them read the words aloud. Point out that when we write how many hundreds we have, as in Task 1(b) on p. 15, we use the plural *hundreds*. So, for 346, we have 3 hundreds, 4 tens, and 6 ones. But when we say the number, *three hundred forty-six*, we do not use the plural form.	346 3 hundreds 4 tens 6 ones Three hundred forty-six
Write **450** and **405** and have students read the numbers. Write the number words. Point out that in 450, the 5 is the number of tens. We have 5 tens, or *fifty*. So we say and write the number as *four hundred fifty*. In 405, we have no tens, so we say and write *four hundred five*.	450 Four hundred fifty 405 Four hundred five
Assessment	**Text p. 19**
Have students write the numbers for these tasks in words rather than figures.	7. six hundred twenty-three 8. (a) four hundred sixty-seven (b) two hundred fifty (c) three hundred six
Reinforcement	
Write the problems shown here and get students to supply the missing numbers.	50 + 5 + 200 = ______ (255) 500 + 4 = ______ (504) 900 + ______ + 5 = 935 (30) 800 + 30 = ______ (830) 3 + 80 + ______ = 383 (300) 100 + 15 = ______ (115) 900 + ______ = 991 (91) 1 + 40 + 300 = ______ (341) 60 + 2 + 100 = ______ (162) ______ + 400 = 405 (5)

Write the problems shown here or similar ones and get students to write the numbers. You can have them also write the numbers in words. Allow students to use a place-value chart and discs, if needed, renaming where necessary.	7 tens 2 hundreds 4 ones (724) 3 ones 2 hundreds (203) 2 hundreds 14 tens (340) 6 hundreds 1 tens 36 ones (646) 5 hundreds 23 tens 12 ones (742)
Practice	WB Exercise 6, pp. 21-23

Activity

Material: Place-value charts for each student, place-value discs for tens and ones mixed up in a bag or box, place-value discs for hundreds.

Procedure: Have students pick out a handful of discs and place them on their charts. There may be too many ones or tens. Get them to trade in 10 ones for a ten and 10 tens for a hundred if needed and show the number on the place-value chart as it would be written.

Tell students that a place on the chart can hold more than 9 discs, and the discs represent the same number whether there are, for example, 23 ten-discs or 2 hundred-discs and 3 ten-discs, or even 230 one-discs. But in order for the number we write to match the discs on the chart, any time there are more than 9 in a column, we trade in 10 ones for a ten and 10 tens for a hundred.

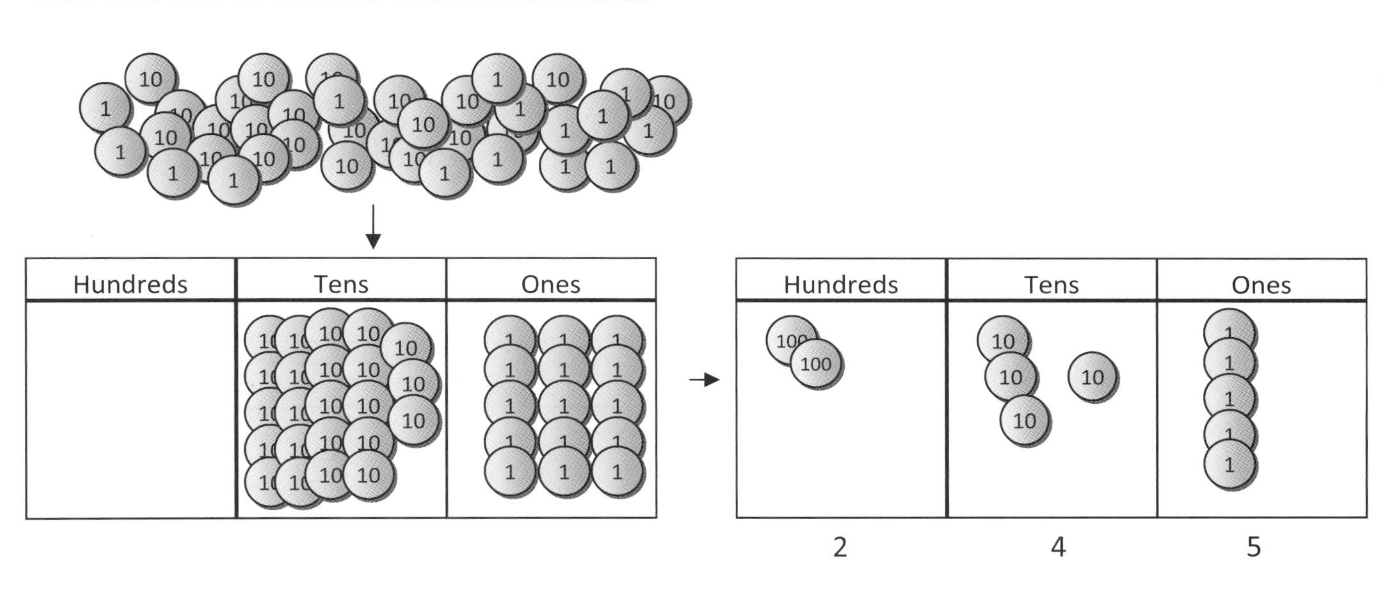

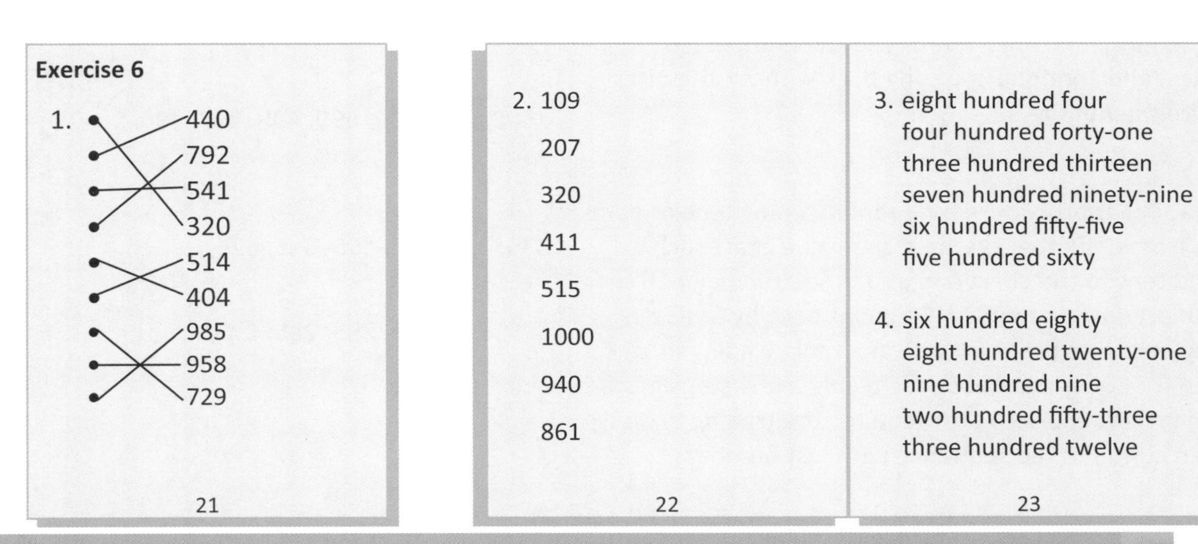
Exercise 6

1. 440, 792, 541, 320, 514, 404, 985, 958, 729

21

2. 109
207
320
411
515
1000
940
861

22

3. eight hundred four
four hundred forty-one
three hundred thirteen
seven hundred ninety-nine
six hundred fifty-five
five hundred sixty

4. six hundred eighty
eight hundred twenty-one
nine hundred nine
two hundred fifty-three
three hundred twelve

23

1.3d Count On and Count Back

Objectives

- Count on 1, 10, or 100.
- Count back 1, 10 or 100.
- Count on by ones, tens, or hundreds.
- Count back by ones, tens, or hundreds.

Note

For more capable students include counting on or back by 2, 3, 20, 30, 200, and 300 and have students do Mental Math 10.

Read and write number words	
Provide students with place-value charts and discs, or demonstrate on the board. Write the number **498** and have students represent the number with the discs. Then ask them to show you the number that is 1 more the 498. They should add a one to the ones column. Ask them again for the number that is one more. They add another one, but now they have to trade in the 10 ones for a ten, and the 10 tens for a hundred. Continue to 502. Write the numbers, have them read the numbers in the sequence, and continue counting on without the discs.	Hundreds / Tens / Ones 4 / 9 / 8 498, **499, 500, 501, 502**, 503, 504, ...
Now ask students to show you the number that is one less than 502. Continue asking them to show you the number that is one less for a few more times. To show one less than 500, they have to trade in a hundred for 10 tens, and then a ten for 10 ones. Write the numbers, have them read the numbers in sequence, and continue counting back without the discs as you write the numbers.	Hundreds / Tens / Ones 5 / 0 / 2 502, **501, 500, 499, 498,** 497, 496, ...
Follow a similar procedure for counting on by tens or back by tens from a number on the place-value chart such as 275. Add tens to the chart one at a time, trading in 10 tens for a hundred when needed. To count back by tens, remove tens from the chart one at a time, trading in 1 hundred for 10 tens as needed. Do only a few steps on the chart to include the step that requires the trading in discs, and then continue the sequence with just numbers.	275, 285, 295, 305, 315, 325, ... 325, 315, 305, 295, 285, 275, ...

Assessment	Text p. 20
Task 9: No renaming is needed for these. Task 10: These involve renaming, but students should be able to do these by simply counting on or back. You can have them use place-value discs if needed, and then provide additional practice until they can count on or back 1, 10, or 100 without discs.	9. (a) 254 (b) 133 (c) 241 10. (a) 800 (b) 490 (c) 570
Reinforcement	
Have students count orally on or back from a given number by ones, tens, or hundreds. Write a starting number and give students a rule to count on or back in steps of 1, 10, or 100. Have students write the next 5 numbers down.	786: Count on in steps of 10. 796, 806, 816, 826, 836 321: Count back in steps of 10. 311, 301, 291, 281, 271 206: Count on in steps of 100. 206, 306, 406, 506, 606
Enrichment	
Write a starting number and give students a rule to count on or back in steps of 2, 3, 20 or 30.	
Write two numbers within 100 and ask students to count from one to the other, first by hundreds, then by tens, and then by ones.	248, 572 Forwards: 248, 348, 448, 548, 558, 568, 569, 570, 571, 572 Backwards: 572, 472, 372, 272, 262, 252, 251, 250, 249, 248
Write some addition problems involving adding or subtracting 1, 2, 3, 10, 20, or 30 to or from a 3-digit number.	348 + 2 871 – 3
Mental math practice	Mental Math 10
Practice	WB Exercise 7 p. 24

Exercise 7

1. (a) 335
   (b) 420
   (c) 506

2. (a) 573
   (b) 774
   (c) 508
   (d) 840

24

1.3e Compare and Order Numbers within 1000

Objectives

- Compare and order numbers within 1000.
- Practice.

Note

Some students may initially have difficulty understanding why a number represented by more discs is smaller than a number represented by fewer discs. If this is the case have students represent the numbers with base-10 blocks instead of place-value discs initially so they can visually see that 465, for example, is smaller than 540, even though more discs are used to represent the number. Then have them represent the same numbers with place-value discs and have them focus on the number of discs in each place.

Compare two 3-digit numbers	
Provide students with place-value charts and place-value discs. Write **465** and **540** and ask students to illustrate the numbers with place-value discs. Then ask students to tell you which number is greater and why it is greater. 540 is greater than 465 because it has more hundreds.	Hundreds / Tens / Ones 100 100 100 100 / 10 10 10 10 10 10 / 1 1 1 1 1 4 / 6 / 5 100 100 100 100 100 / 10 10 10 10 / 5 / 4 / 0 540 is greater than 465 because it has more hundreds.
Repeat with **234** and **267**. Point out that the hundreds are the same, so we compare the tens. 267 is greater than 234 because it has more tens.	267 is greater than 234 because it has more tens.
Repeat with **123** and **125**. Tell students we first compare the hundreds. If they are the same, we then compare the tens. If they are the same, we then compare the ones. 125 is greater than 123 because it has more ones.	125 is greater than 123 because it has more ones.
Write two 3-digit numbers, such as **565** and **546**, one above the other with the digits aligned, and have students tell you which one is larger without using place-value discs. Ask students to write the numbers side by side and put the correct symbol between them.	5 6 5 5 4 6 565 > 546

Assessment	
Write the numbers below on the board. For the first three, ask students for the correct symbol, > or <, that should go between the numbers.	
⇒ 348 601	348 < 601
⇒ 821 97	821 > 97
⇒ 362 368	362 < 368
⇒ Which number is smallest: 241, 421, or 214?	214 is smallest
⇒ Which number is greatest: 645, 654, or 456?	654 is greatest
⇒ What is the greatest number that can be made with 1, 2, and 3?	321
⇒ What are the smallest and greatest 3 digit numbers?	100 is smallest, 999 is greatest
Enrichment	
Write **400 + 30** next to **403**. Ask students to put the symbol ">" or "<" between them. Tell them that the symbol ">" means that the total on the left is greater than the total on the right. To determine what symbol to use, we have to first find the value for 400 + 30. Write some additional problems for students to solve.	400 + 30 > 403 65 tens ( ) 65 (>) 500 + 3 + 30 ( ) 3 + 500 + 30 (=) 4 hundreds 8 ones ( ) 480 ones (<)
Write the equations or inequalities at the right and ask students to fill in the blanks with a number that will make them true.	4 + 5 > 5 + _____ (3, 2, 1, or 0) 8 + 2 = 9 + _____ (1) 6 – 4 < 6 – _____ (3, 2, 1, or 0)
Practice	**Text p. 21, Practice 1B**
Class game Material: Cards with numbers within 1000, enough for each student to have one. Procedure: Divide students into teams. Give each team member a number card. Students must line up in the order of the numbers on their cards. The first team to get in the proper order wins. **Group game** Material: Number cube 1-6, 1 per group. Procedure: Students draw three lines or a place-value chart for tens and ones on their paper. Each student rolls the number cube once, decides whether the number should be a hundred, a ten, or a one, and writes it in the appropriate place. Repeat two more times to fill in the remaining places. The student with the greatest 3-digit number at the end of a round gets a point. Students can also record all of their group's numbers in order for each round.	1. (a) three hundred thirty (b) one hundred forty-four (c) two hundred fifty-five (d) six hundred eight 2. (a) 6 hundreds 4 tens 5 ones (b) 7 hundreds 2 tens 0 ones (c) 4 hundreds 0 tens 9 ones (d) 9 hundreds 3. (a) 704 (b) 540 (c) 304 (d) 820 4. (a) > (b) < (c) > (d) > 5. (a) 99, 410, 609 (b) 104, 140, 401, 410 6. (a) 300 (b) 779 7. (a) 472 (b) 790 8. (a) 699 (b) 505

2 Addition and Subtraction

Objectives

- Review the part-whole concept of addition and subtraction.
- Compare the quantities in two sets by using subtraction.
- Add and subtract numbers within 1000.
- Solve word problems involving addition and subtraction within 1000.

Suggested number of weeks: 6

		TB: Textbook WB: Workbook	Objectives	Material	Appendix
2.1	**Meanings of Addition and Subtraction**				
2.1a	Review: Part-Whole	TB: pp. 22-24 WB: p. 25	♦ Review the part-whole concept of addition and subtraction.	♦ Counters	a17
2.1b	Review: Difference	TB: p. 24 WB: pp. 26-27	♦ Compare the quantities in two sets by using subtraction.	♦ Counters	a18
2.1c	Review: Mental Math	TB: p. 25 WB: pp. 31, 36	♦ Review addition and subtraction of 2-digit numbers, no renaming.	♦ Place-value discs and charts	Mental Math 11
2.1d	Word Problems	TB: pp. 26-27, 34 WB: pp. 28-29	♦ Solve word problems involving addition and subtraction.	♦ Place-value discs and charts	a19
2.2	**Addition Without Renaming**				
2.2a	Add a 2-digit Number	TB: p. 29 WB: pp. 30, 32	♦ Add a 2-digit number to a number within 1000, no renaming.	♦ Place-value discs and charts	
2.2b	Add a 3-digit Number	TB: pp. 28, 30 WB: pp. 33-34	♦ Add a 3-digit number to a number within 1000, no renaming.	♦ Place-value discs and charts ♦ Number cards	Mental Math 12
2.3	**Subtraction Without Renaming**				
2.3a	Subtract a 2-digit Number	TB: p. 32 WB: pp. 35, 37	♦ Subtract a 2-digit number from a number within 1000, no renaming.	♦ Place-value discs and charts	
2.3b	Subtract a 3-digit Number	TB: pp. 31, 33 WB: pp. 38-39	♦ Subtract a 3-digit number from a number within 1000, no renaming.	♦ Place-value discs and charts	Mental Math 13
2.3c	Practice	TB: p. 35	♦ Add and subtract within 1000, no renaming. ♦ Solve word problems involving addition and subtraction.	♦ Number cards	Mental Math 14

		TB: Textbook WB: Workbook	Objectives	Material	Appendix
2.4	**Addition With Renaming**				
2.4a	Add Ones or Tens to a 3-digit Number	TB: p. 37 WB: p. 40	♦ Add ones or tens to a 3-digit number using mental math strategies.	♦ Place-value discs and charts	Mental Math 15
2.4b	Rename Ones	TB: pp. 37-38 WB: pp. 41-42	♦ Add within 1000 by renaming ones.	♦ Place-value discs and charts	
2.4c	Rename Tens	TB: pp. 38-39 WB: p. 43	♦ Add within 1000 by renaming tens.	♦ Place-value discs and charts ♦ Number cubes	
2.4d	Word Problems	TB: pp. 41-42 WB: pp. 44-45	♦ Add within 1000 by renaming ones or tens. ♦ Solve word problems.	♦ Place-value discs and charts	
2.4e	Rename Tens and Ones	TB: pp. 39-40 WB: pp. 46-47	♦ Add within 1000 by renaming both tens and ones.	♦ Place-value discs and charts ♦ Number cards	
2.4f	Word Problems	TB: pp. 40-42 WB: pp. 48-49	♦ Add three numbers. ♦ Solve word problems.		
2.5	**Subtraction With Renaming**				
2.5a	Subtract Ones or Tens from a 3-digit Number	TB: p. 44	♦ Subtract ones or tens from a 3-digit number using mental math strategies.	♦ Place-value discs and charts	Mental Math 16-17
2.5b	Rename Tens	TB: pp. 44-45 WB: p. 51	♦ Subtract within 1000 by renaming tens.	♦ Place-value discs and charts ♦ Base-10 blocks	
2.5c	Rename Hundreds	TB: pp. 45-46 WB: p. 54	♦ Subtract within 1000 by renaming hundreds.	♦ Place-value discs and charts ♦ Number cubes	
2.5d	Word Problems	TB: p. 48 WB: pp. 52-53, 55	♦ Subtract within 1000 by renaming tens or hundreds. ♦ Solve word problems.		a20
2.5e	Rename Tens and Hundreds	TB: pp. 46-47 WB: pp. 56-57	♦ Subtract within 1000 by renaming both tens and hundreds.	♦ Place-value discs and charts	
2.5f	More Subtraction	TB: pp. 47, 49 WB: pp. 58-60	♦ Subtract within 1000 when there are no tens. ♦ Solve word problems.	♦ Place-value discs and charts ♦ Base-10 blocks	
2.5g	Practice and Review	TB: pp. 50-51 WB: pp. 61-66	♦ Practice addition and subtraction within 1000. ♦ Review all topics.	♦ Number cards	Mental Math 18

2.1 Meanings of Addition and Subtraction

Objectives

- Review the part-whole concept of addition and subtraction.
- Compare the quantities in two sets by using subtraction.
- Relate "more than" and "less than" to subtraction.
- Review addition and subtraction of 2-digit numbers, no renaming.
- Solve word problems involving addition and subtraction.

Material

- Counters for students and board
- Place-value discs and charts for students and board
- Mental Math 11 (appendix)
- Appendix pp. a17-19

Prerequisites

Students should know how to mentally add or subtract a 1-digit number to or from a 2-digit number and add 2-digit numbers without renaming. This will be reviewed, but you can also go back to Units 3 and 8 in *Primary Mathematics* 1B for more systematic instruction and practice.

Notes

In *Primary Mathematics* 1A students learned to associate addition and subtraction with the part-whole concept and number bonds.

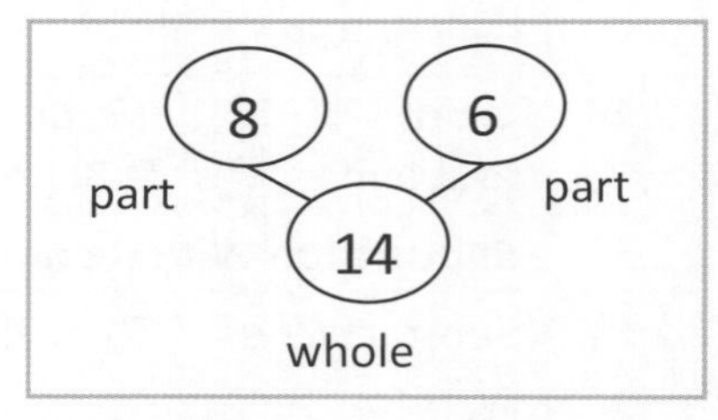

A number bond simply shows two parts linked to a whole. It can be represented in a number of ways, with a shape around the number or not, and in any orientation. When drawing number bonds to help in solving word problems, using circles or squares and writing the number in them helps to distinguish the parts and the whole.

To find the whole when given two parts, we add.

To find a missing part, when given one part and the whole, we subtract.

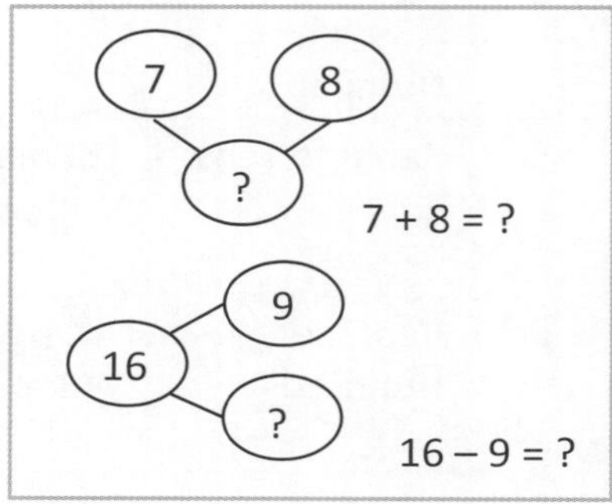

When interpreting word problems, students can use the part-whole concept to determine whether to add or subtract to find the answer. They should read the problem carefully and decide whether it gives two parts and is asking for a total (e.g., Tasks 9 and 10, p. 26 in the textbook) or gives a total and one part and is asking for the missing part (e.g., Task 11, p. 27 in the textbook). Students can draw an empty number bond and fill it in with information from the problem to help them determine what equation to use.

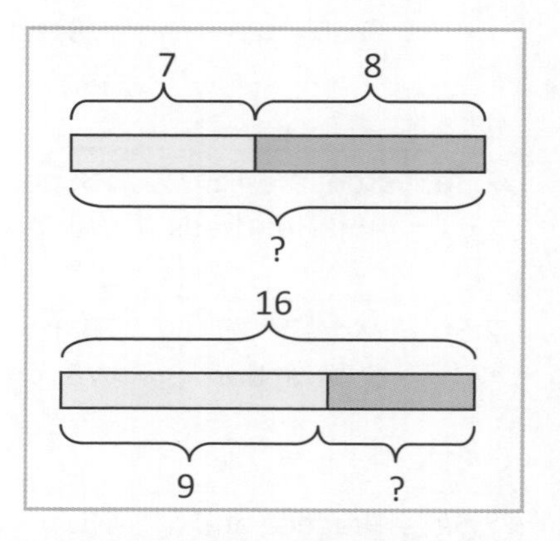

In *Primary Mathematics* 3A, students will learn to draw bar models to represent the information in the problem. Although bar models are introduced early in some of the supplementary practice books, they are not introduced until the 3rd grade in *Primary Mathematics*, where they are more appropriate. At this level, students should focus on determining whether the problem gives the parts or is missing a part, a skill that is needed to even draw a model. Number bonds are sufficient, if needed, and are easier to draw, or the student can act out the situation. Once it is determined whether there is a missing part or not for these types of simple problems, the solution is evident, so drawing the model becomes redundant and distracting. Bar models are more useful for multi-step problems, by which time students can draw them neatly enough to then get information from them as well as make the bars proportional. You can still draw bar models yourself to help explain the problem to students, and have them tell you what numbers to write, but do not require students to draw bar models yet, unless you intend to teach them to solve the more challenging multi-step problems from supplemental books.

In *Primary Mathematics* 1B, students learned to use subtraction to compare two numbers to find out how many more or less there are in one set than in another.

For example, there are 10 ice cream cones and 6 ice cream bars.

To find out how many more ice-cream cones there are than ice-cream bars (or how many fewer ice-cream bars than ice-cream cones), we subtract: 10 – 6 = 4.

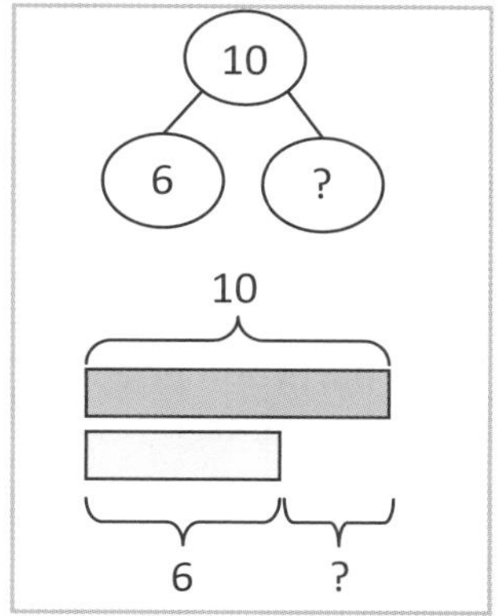

In solving word problems involving comparison, students can determine if the problem asks whether there are more or less of one thing than another, and then determine which number is larger. This can be illustrated with a number bond where the smaller number is a part, the larger number is a whole, and the difference is the missing part.

You can also illustrate the problem with a comparison model, but do not require students to draw the models themselves until *Primary Mathematics* 3. With simple one-step problems, once they have interpreted the problem sufficiently to even draw the model, they should know how to find the answer, and drawing precise models at this age is time-consuming and redundant for these types of problems.

Do not have students simply look for key words such as "more" or "altogether" in determining whether the problem involves comparison or addition. The word problems in *Primary Mathematics* do not follow a pattern in how they are worded. A problem might be: "There are 10 pieces of candy in the jar altogether. Sarah ate 2 pieces of candy. How many pieces are still in the jar?" A student that automatically adds the two numbers he or she sees because the problem includes the word "altogether" will get an incorrect answer. In a later lesson, students will be deciphering two word problems that have similar wording, but are not quite the same. One is solved with addition, the other subtraction (problems 6 and 7 on p. 48).

Students who have difficulty with word problems can act them out, a far more useful strategy at this level than bar models, using counters or place-value discs. If the problem involves larger numbers, you can substitute smaller numbers, guide the student in acting out the problem with counters, draw a number bond to represent the problem if needed, and then substitute in the larger numbers of the original problem.

All the problems in the first part of the unit which involve numbers past 20 do not require renaming. Students should be able to do all the addition and subtraction of 2-digit numbers mentally by simply adding or subtracting the tens and the ones separately, as shown with number discs on pp. 26-27 in the textbook.

Students can also first add or subtract the tens to get an intermediate answer, and then add or subtract the ones, as was taught in *Primary Mathematics* 1B. This strategy will be reviewed in *Primary Mathematics* 2B.

$$43 + 25 = ? \qquad 43 \xrightarrow{+20} 63 \xrightarrow{+5} 68 \qquad 43 + 25 = 68$$

$$58 - 32 = ? \qquad 58 \xrightarrow{-30} 28 \xrightarrow{-2} 26 \qquad 58 - 32 = 26$$

When discussing the word problems, have students read them out loud. Also have them read or state an answer sentence out loud, including the answer. The answer sentence can be simple, but using a sentence will help students determine if the numerical value they obtained actually answers the question posed in the problem. Do not have them write the answer sentence with a blank before finding the answer, in which case they might just fill in the blank without rereading the sentence to be sure they answered the question.

When students solve problems you have written on the board or from the textbook during a lesson, they can work out the answers on their whiteboards, and then hold them up when they are done.

2.1a Review: Part-Whole

Objectives

- Review the part-whole concept of addition and subtraction.
- Use addition to find the whole.
- Use subtraction to find the part.
- Interpret word problems for whole or parts.

Vocabulary

- Number bond
- Part
- Whole
- Equation

Review part-whole concept	
Show students two sets of counters or draw them on the board. Draw the number bond using a question mark for the whole. Call it a *number bond*. Tell them that you have two *parts*. Bring both sets together and tell them the two parts make a *whole*, which is the total amount. Ask students how we show the process of putting two parts together with a mathematical sentence. Write two addition equations. Tell them that when we know the amount in both parts and need to find the total, or whole amount, we add. It does not matter in what order we write the numbers for the two parts.	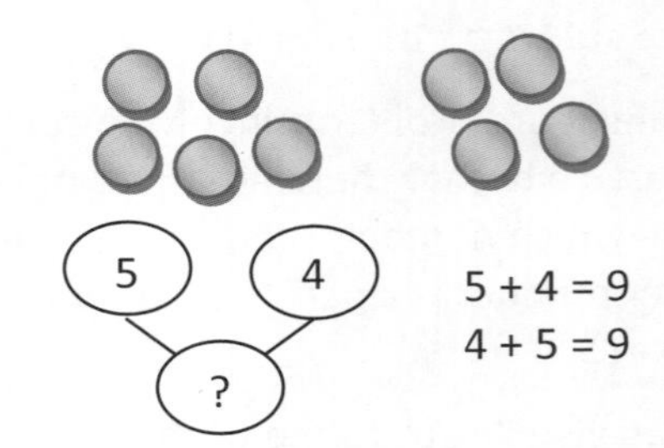
Tell students that these "number sentences" are called *equations*. The equation shows that the two parts are equal to the whole. The values on both sides of the "equal" sign are the same, but one side shows the two parts and the other side shows the whole. We can write an equation where one side is not a single number, such as 5 + 4 = 4 + 5, and the equation is still true, because both sides are the same as 9. 9 = 4 + 5 is also true.	5 + 4 = 4 + 5 9 = 4 + 5
Draw another number bond and write the whole in it. Cover up some of the counters or draw a container on the board representing the hidden counters. Ask students how many are showing and write that number as a part in the number bond. Ask them how many are covered up or are in the container. Ask them what kind of an equation we would write to show that we need to find a part we don't know. Write the subtraction equation. Repeat, this time covering up the other part.	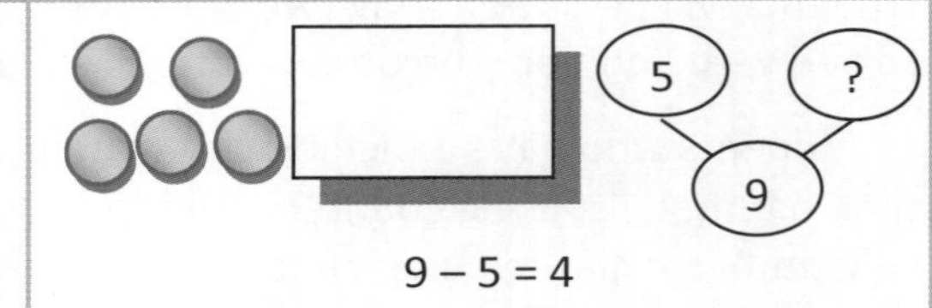 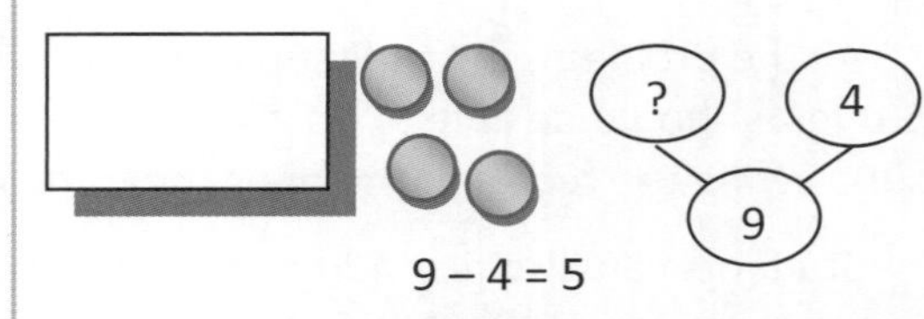
Write **9 – 5 = 9 – 4** and ask if it is true. It is not. The two sides of the equation are different. In a subtraction equation, we always write the whole first.	~~9 – 4 = 9 – 5~~
Discussion	**Text pp. 22-24**
Have students read the problem on p. 22 in the textbook out loud. Ask whether the problem is asking us to find the whole or whether it gives us a whole. Have students identify the parts and supply the answers. Have them read the answer sentence out loud.	14; 14
Task 1: Have students read the problem out loud. Ask whether it is asking us to find the whole or whether it gives us the whole. Have students identify the part that is given and what part we need to find.	1. 6; 6

Task 2: We can use four equations to describe the situation given in the diagram, two addition equations and two subtraction equations. You can ask students make up three different story problems for the situation in Task 2, one that can be solved by either of the addition equations, and one each for the two subtraction equations. Task 3: Have students supply the answers.	2. 7 + 5 = **12** 5 + 7 = **12** 12 – 5 = **7** 12 – 7 = **5** 3. 9 + 3 = **12** 3 + 9 = **12** 12 – 3 = **9** 12 – 9 = **3**
Assessment	Appendix p. a17
Have students do the problems on appendix p. a17. You may have to help some read the problems. They can draw number bonds if needed and fill them in as they determine what information is provided, or act out the problem with counters.	
1. Harriet has 9 parakeets and 5 parrots. How many birds does she have?	9 + 5 = 14 She has 14 birds.
2. The noise from the birds was just too much and so she gave 8 birds away. How many birds does she have left?	14 – 8 = 6 She has 6 birds left.
3. A friend gave her 2 macaws. How many birds does she have now?	6 + 2 = 8 She now has 8 birds.
4. Her landlord complained about the noise and she had to give some birds away again. Now she just has 2 quiet parakeets left. How many birds did she have to give away the second time?	8 – 2 = 6 She gave away 6 birds.
Practice	WB Exercise 8, p. 25

Exercise 8

1. (a) 10, 10, 8, 2 (b) 11, 11, 7, 4
(c) 13, 13, 7, 6 (d) 14, 14, 6, 8
(e) 17, 17, 14, 3 (f) 19, 19, 9, 10
(g) 25, 25, 20, 5 (h) 19, 19, 13, 6

25

2.1b Review: Difference

Objectives

- Compare the quantities in two sets by using subtraction.
- Relate "more than" and "less than" to subtraction.

Vocabulary

- Compare
- Difference
- Less than
- More than

Note

As with the previous lesson, this lesson is primarily review. However, the concepts of part-whole and of finding the difference are central themes to this curriculum, so be sure that students thoroughly understand them.

Compare two quantities	
Display two sets of counters, each set a different color, or draw them on the board. Ask students which set has more. Then, line up the counters. Tell students that we are *comparing* the amount in the two sets by finding the *difference*. Point out that we can think of the smaller set as a part, the larger set as the whole, and the difference as a missing part. Ask them what equation we can write to answer the question "How many more are in the set of 9 than in the set of 6?" Write the subtraction equation. Ask them what equation we would write to answer the question "How many less are in the set of 6 than in the set of 9?" We would write the same equation. Show students that we can represent the problem with a number bond. You can make the circle for the larger number slightly more oval to relate the number bond to the lined-up disks.	$9 - 6 = 3$ 9 6 3
Write the problems shown at the right and have students supply the answers that would go in the blanks.	6 less than 9 is _____. _____ is 6 less than 9. 9 is _____ more than 6. _____ more than 6 is 9. _____ less than 9 is 6. 6 more than 9 is ____.
Discussion	**Text p. 24**
Task 4(a): Students can find the answer without counting by simply covering up the same number in B as in A. The remaining number of bananas is the answer. Task 4(b): Students should count the number in each set. Task 5: Also ask: 14 is how many more than 8?	4. (a) 4 (b) 9 – 5 = **4** 5. (a) 14 – 8 = **6** (b) 6

Assessment	Appendix p. a18
Have students tell you why these are comparison problems and which quantities are being compared. They can draw number bonds if needed, using a question mark to indicate what needs to be found, or act the problems out with counters. They should then write an equation. Have them supply answers in a complete sentence. Students may use counters if needed to act out the problems.	
1. Harriet started out with 9 parakeets and 5 parrots. How many more parakeets than parrots did she have?	9 – 5 = 4 There were 4 more parakeets than parrots. (9: 5, ?)
2. After giving some birds away, Harriet ended up with 2 parakeets. She was finally able to move out of an apartment and get a large house. She now has 20 birds. How many more birds does she have now than she had before?	20 – 2 = 18 She has 18 more birds. (2, ?: 20)
3. Harriet saw a blue parakeet at the pet store. It cost $18. She had $6 with her. How much more money does she need?	$18 – $6 = $12 She needs $12 more. ($18: $6, ?)
Reinforcement	
Ask students to make up some story problems involving the comparison of two quantities.	
Practice	WB Exercise 9, pp. 26-27

Exercise 9

1. (a) 5
   5
   (b) 7
   7
   (c) 6
   6

26

2. (a) 8
   8
   (b) 7
   7
   (c) 9
   9
   (d) 8
   8
   (e) 4
   4
   (f) 8
   8

27

2.1c Review: Mental Math

Objectives

- Review addition and subtraction of 2-digit numbers, no renaming.

Note

In order to answer the word problems in Exercise 10, students must be able to add 2-digit numbers where there is no renaming. Students learned to do this mentally in *Primary Mathematics* 1B. This is covered in the next unit briefly before going into addition and subtraction of 3-digit numbers. The problems from those exercises are therefore going to be assigned for this lesson.

For each example problem shown here, write the addition or subtraction expression, show the discs on the chart, and either show that we add tens to tens and ones to ones, or subtract tens from tens and ones from ones. Use the place-value of the digit every time you refer to it. For example, say, "We add 4 *tens* and 5 *tens* to get 9 *tens*" rather than, "We add 4 and 5 to get 9." You may wish to show the process with number bonds as shown here. However, do not require students to draw number bonds themselves; they should be able to do these types of problems mentally. Then, for each type of problem, have students solve the other problems with or without the discs. Since this is a review, more capable students will not need discs at all.

Add or subtract	Tens	Ones		
Add tens together. 4 tens + 5 tens = 9 tens	10 10 10 10 10 10 10 10 10		40 + 50 = ? 40 + 50 = 90 10 + 80 20 + 30 30 + 50	 (90) (50) (80)
Add tens to a 2-digit number. 3 tens 4 ones + 2 tens 34 + 20 = 54 4 30	10 10 10 10 10	1 1 1 1	34 + 20 = ? 34 + 20 = 54 22 + 40 79 + 10 47 + 30	 (62) (89) (77)
Add ones to a 2-digit number. 3 tens 4 ones + 5 ones 34 + 5 = 39 30 4	10 10 10	1 1 1 1 1 1 1 1 1	34 + 5 = ? 34 + 5 = 39 23 + 6 51 + 4 92 + 7	 (29) (55) (99)
Add tens and ones to a 2-digit number. 3 tens 4 ones + 2 tens 5 ones 34 + 25 = 59 30 4 20 5	10 10 10 10 10	1 1 1 1 1 1 1 1 1	34 + 25 = ? 34 + 25 = 59 55 + 23 23 + 55 25 + 53	 (78) (78) (78)

Subtract tens. 7 tens – 5 tens	Tens: 10 10 10 10 10 (crossed out), 10 10 | Ones:	70 – 50 = ? 70 – 50 = 20 90 – 60 (30) 80 – 20 (60) 40 – 30 (10)
Subtract tens from a 2-digit number. 7 tens 6 ones – 5 tens 76 – 50 = 26 6 70	Tens: 10 10 10 10 10 (crossed out), 10 10 | Ones: 1 1 1 1 1, 1	76 – 50 = ? 76 – 50 = 26 84 – 50 (34) 76 – 50 (26) 55 – 20 (35)
Subtract ones from a 2-digit number. 7 tens 6 ones – 2 ones 76 – 2 = 74 70 6	Tens: 10 10 10 10 10, 10 10 | Ones: 1 1 1 1 1, 1 (two crossed out)	76 – 2 = ? 76 – 2 = 74 27 – 5 (22) 48 – 3 (45) 99 – 2 (97)
Subtract tens and ones from a 2-digit number. 7 tens 6 ones – 5 tens 2 ones 76 – 52 = 24 70 6 50 2	Tens: 10 10 10 10 10 (crossed out), 10 10 | Ones: 1 1 1 1 1, 1 (two crossed out)	76 – 52 = ? 76 – 52 = 24 78 – 23 (55) 78 – 55 (23) 49 – 33 (16)
Assessment	**Text p. 25**	
Task 8: You can ask students to draw a number bond for this problem.	6. 56 7. 14 8. 32 + 13 = **45** 45 – 13 = **32** 13 + 32 = **45** 45 – 32 = **13** (number bond: 32, 13 → 45)	
Mental math practice	Mental Math 11	
Practice	WB Exercise 11, p. 31, problem 3 WB Exercise 13, p. 36, problem 3	

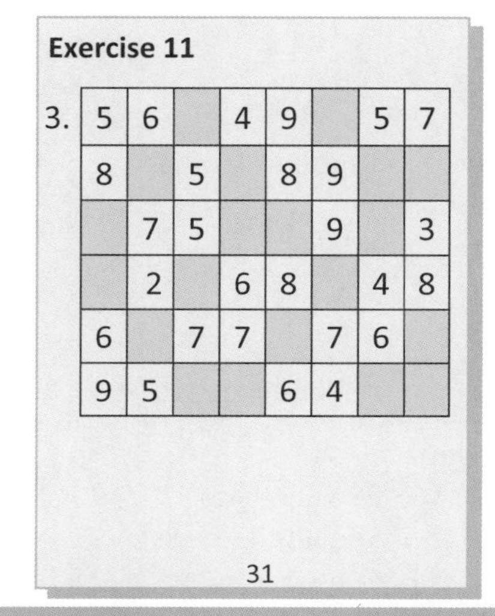

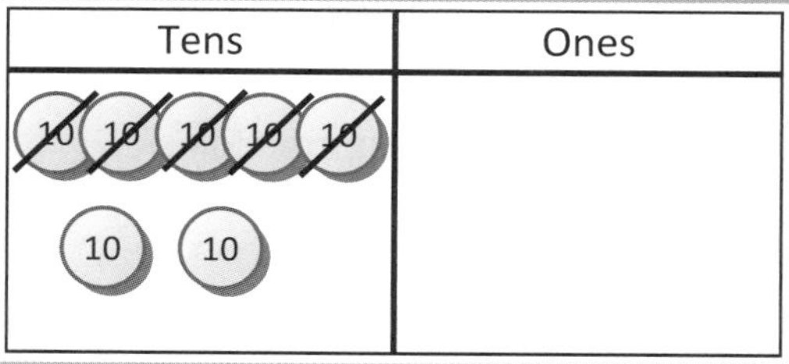

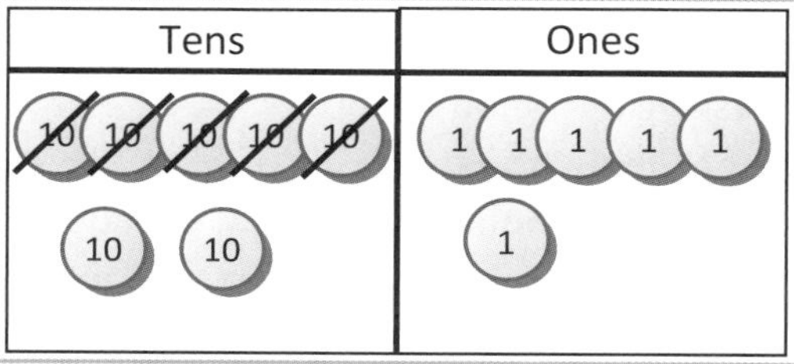

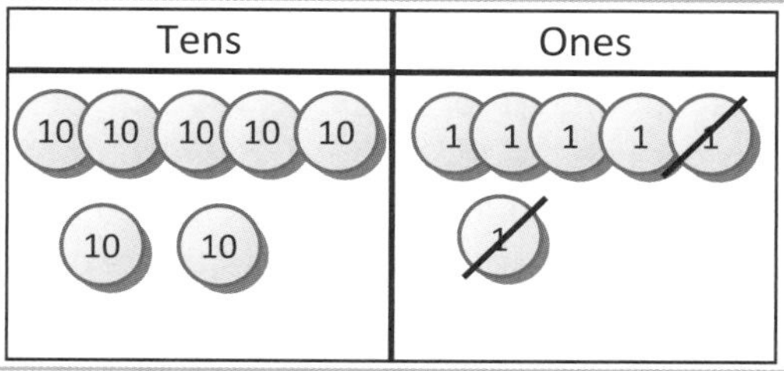

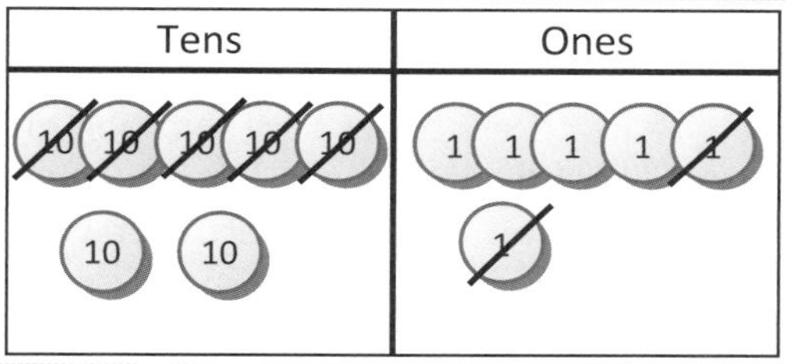

Exercise 11

3.

5	6		4	9		5	7
8		5		8	9		
	7	5			9		3
	2		6	8		4	8
6		7	7		7	6	
9	5			6	4		

31

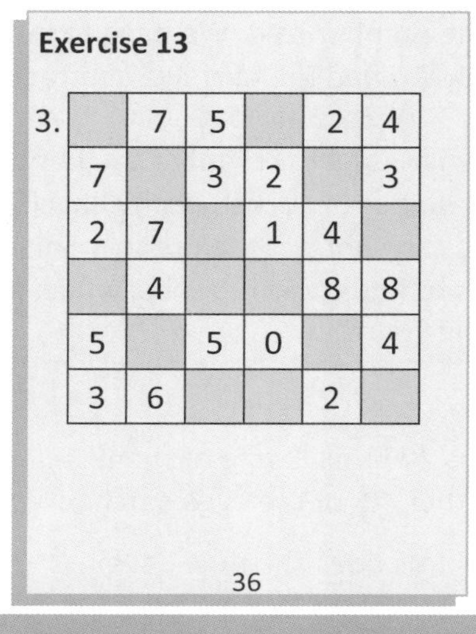

Exercise 13

3.

	7	5		2	4
7		3	2		3
2	7		1	4	
	4			8	8
5		5	0		4
3	6			2	

36

2.1d Word Problems

Objectives

- Solve word problems involving addition and subtraction.

Note

For each of the problems in the textbook, ask students to read the problem out loud. Then ask them questions such as the following.

- ⇒ What is the problem asking us to find?
- ⇒ Are we trying to find the whole (the total)? If so, what are the parts?
- ⇒ Does the problem give a total amount? If so, what part does it give? What part do we have to find?
- ⇒ Are we comparing two amounts? If so, which two amounts? Which is the larger amount and which is the smaller amount? What do we need to find? We can write the larger amount for the whole, and the smaller amount for a part. How do we find the difference?

If possible, rather than having students refer to the textbook, where the equation is already given, write out the problems on the board and guide students through the solution. Draw a number bond (or a bar model if you prefer, but don't require students to draw the models) and fill in the parts as you discuss the problem. If instead you are having students read the problems from the textbook, then for Tasks 9 and 10 you can ask them why addition was used, and for Tasks 11 and 12 you can ask why subtraction was used. If you are having students draw number bonds, they can just use the numbers, not fill in words as well, the number bonds in this guide are for your reference.

If some students have difficulty, allow them to act out the problems, using place-value discs to represent items in the problem. If they still have difficulty, rephrase the problem with single-digit numbers. After they have found the answer, using objects if needed, repeat the problem with the large numbers.

Discussion	Text pp. 26-27
Task 9: The problem is asking us to find the total number of key chains he now has, which is the whole amount. The two parts are the number of key chains Danny already has and the number that he buys. We add the parts to find the total.	9. 34 + 5 = **39** 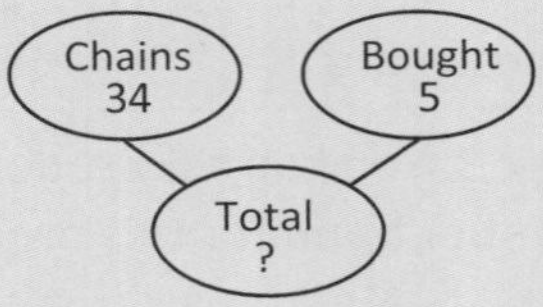 Chains 34, Bought 5 → Total ?
Task 10: The problem is asking us to find the total number of apples, which is the whole amount. It gives us two parts, the number of green apples and the number of red apples. We add to find the total.	10. 24 + 32 = **56** 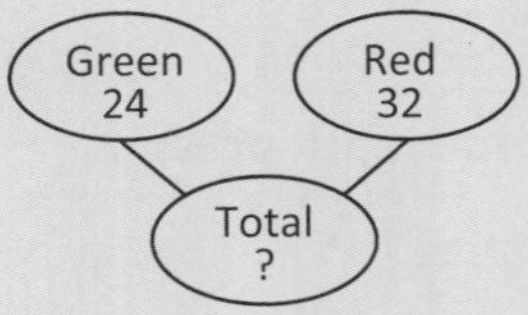 Green 24, Red 32 → Total ?
Task 11: The problem gives us the total number of goldfish. It then gives us one part, the number sold. We have to find the other part, the number left. To find the part, we subtract.	11. 78 – 40 = **38** 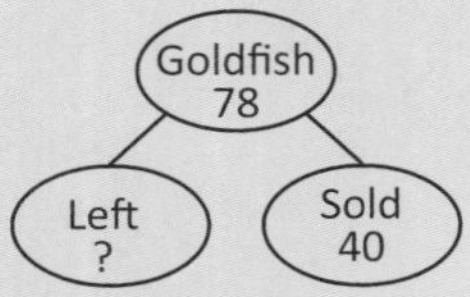 Goldfish 78 → Left ?, Sold 40
Task 12: The problem asks us to compare the number of stickers Rahmat has to the number of stickers Samy has. Rahmat has the most stickers. So we can subtract the number Samy has from the number Rahmat has to find the difference, which is how many more Rahmat has than Samy.	12. 48 – 32 = **16** 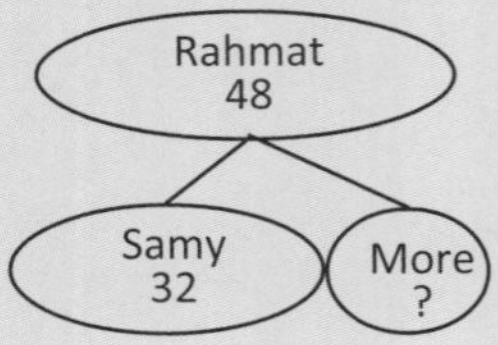 Rahmat 48 → Samy 32, More ?
Assessment	Appendix p. a19
1. Harriet now has 20 birds. Some of them laid eggs. There were 37 eggs altogether. 25 of the eggs hatched. (a) How many birds does she now have?	20 + 25 = 45 She has 45 birds. 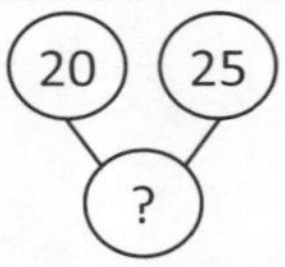 20, 25 → ?

(a) How many eggs did not hatch?

37 – 25 = 12
12 eggs did not hatch.

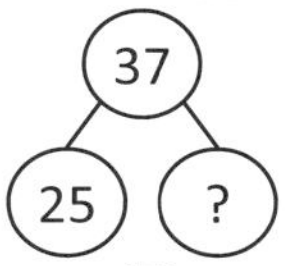

(b) She sold 13 of her birds. How many does she have left?

45 – 13 = 32
She has 32 birds

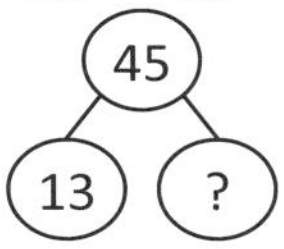

(c) How many more birds does she have now than she had before they started laying eggs and she sold some?

32 – 20 = 12
She now has 12 more birds.

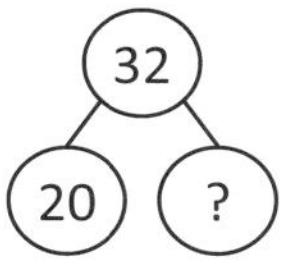

Assessment	**Text p. 34, Practice 2A**
You can ask students to draw number bonds for the problems in the workbook. Do not, however, require number bonds if a student does not need to draw them to solve the problem. Drawing a number bond, and at later levels drawing bar models, are simply tools the student can use if he or she is uncertain of which operation to use to solve the problem. They are not required or necessarily needed for every single problem, and the goal is to get students to not need them all the time.	1. (a) 37 (b) 76 (c) 88 2. (a) 61 (b) 39 (c) 51 3. (a) 77 (b) 99 (c) 98 4. (a) 20 (b) 36 (c) 22 5. (a) 89 (b) 40 (c) 5 6. 36 – 11 = **25** She had 25 stamps left. 7. 43 + 24 = **67** He bought 67 sticks of satay. 8. 48 – 25 = **23** (comparison) There are 23 more cherries than kiwis. 9. 23 + 76 = **99** He sold 99 cans altogether. 10. \$19 – \$14 = **\$5** She needs \$5 more.
Practice	WB Exercise 10, pp. 28-29

Exercise 10

1. 45 – 31 = **14**
   14
2. 37 – 24 = **13**
   13
3. 23 + 24 = **47**
   47

28

4. (a) \$32 + \$2 = \$34
   \$34
   (b) \$23 – \$11 = \$12
   \$12
   (c) \$86 – \$15 = \$71
   \$71
   (d) \$32 – \$30 = \$2
   \$2

29

2.2 Addition Without Renaming

Objectives

- Add a 2-digit number to a number within 1000, no renaming.
- Add a 3-digit number to a number within 1000, no renaming.

Material

- Place value discs and charts
- Number cards 0-9, 4 sets per group
- Mental Math 12 (appendix)

Prerequisites

Students should have a good knowledge of place value and of the basic addition facts within 10.

Notes

Up to this point students have been adding 2-digit numbers in a horizontal format using a variety of mental math strategies, including adding the tens first and then the ones. In this unit the standard algorithm for addition is introduced for numbers of up to three digits.

The word algorithm originally meant the art of calculating by means of nine figures and a zero. Here it is used to mean a procedure for solving a mathematical problem in a finite number of steps that frequently involve repetition of an operation. (Students do not need to know the term "algorithm.")

In the standard algorithm for addition, the problem is worked in a vertical format, with the digits aligned in columns, one column for each place value. A line is drawn under the numbers to be added, separating them from their sum. Digits are added starting with the lowest place value, the ones.

In this part students will be adding numbers using the standard algorithm where renaming does not occur. In Part 4 of this unit they will be adding numbers where renaming does occur.

As students become more proficient with adding 3-digit numbers using the vertical format, they may also be able to solve some of these problems in a horizontal format, particularly if they have a good sense of place value and can align the digits mentally. When renaming does not occur, it is easy to add the numbers starting from the left (highest place value) rather then the right. Students should be allowed to do this if they prefer.

Students do need to understand the standard algorithm as taught here thoroughly, since it is the most useful algorithm for all types of problems. When renaming does occur, there are a variety of mental math strategies that can be used, depending on the type of problem. The standard algorithm, however, can be used with any type of problem and should be the fall-back strategy even for those students who like to use mental math. It is particularly efficient in instances where there is renaming over several places, as will be seen later, adding more than two numbers, or adding larger numbers with more digits, especially when many of the digits are larger, such as 7, 8, or 9.

When students solve problems you have written on the board or from the textbook, they can work out the answers on their whiteboards, and then hold them up when they are done to be checked.

If students have difficulty aligning their digits in the vertical format, they can use small place-value charts, graph paper, or lined paper turned sideways.

Hundreds	Tens	Ones
100 100 100	10 10 10 10 10	1 1 1 1 1 1
100 100 100 100	10 10 10	1 1

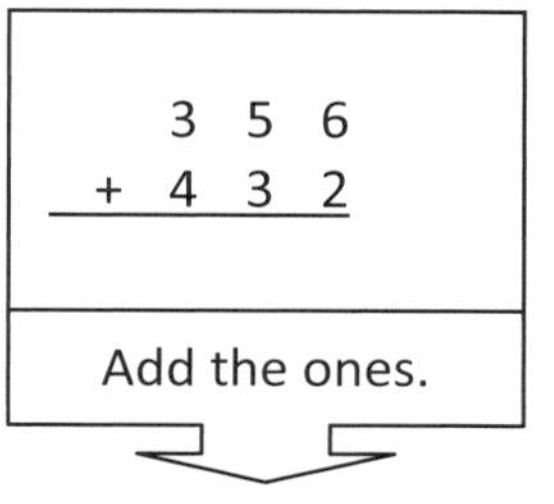

Hundreds	Tens	Ones
100 100 100	10 10 10 10 10	1 1 1 1 1 1 1 1
100 100 100 100	10 10 10	

$$
\begin{array}{r} 356 \\ +\ 432 \\ \hline \mathbf{8} \end{array}
$$

Add the tens.

Hundreds	Tens	Ones
100 100 100	10 10 10 10 10 10 10 10	1 1 1 1 1 1 1 1
100 100 100 100		

$$
\begin{array}{r} 356 \\ +\ 432 \\ \hline \mathbf{8}8 \end{array}
$$

Add the hundreds.

Hundreds	Tens	Ones
100 100 100 100 100 100 100	10 10 10 10 10 10 10 10	1 1 1 1 1 1 1 1

$$
\begin{array}{r} 356 \\ +\ 432 \\ \hline \mathbf{7}88 \end{array}
$$

2.2a Add a 2-digit Number

Objectives

- Add a 2-digit number to a number within 1000 without renaming, using the standard algorithm.

Note

This lesson starts out having students add 2-digit numbers using the addition algorithm. Students with a good understanding of place value will be able to add these mentally. You can tell them that adding ones first will make problems with more digits easier, particularly when the two digits add to more than ten. You might want to write three 3-digit numbers, such as 398 + 476 + 258, and ask if it would be easy to add mentally. Allow students who need them to use place-value charts and discs initially. More advanced students will not need to use place-value charts at all for this lesson.

Add ones, tens, and hundreds	**Text p. 29**
Rewrite the problems vertically. If necessary, illustrate the problems with place-value discs. Tell students that although they know how to add these problems mentally, "in their heads," with more digits it can sometimes be harder to keep track of the place value of each digit. If we write the numbers one above the other, making sure the digits for one place value in the first number line up with the digits for the same place value in the second number, then it is easy to see which digits need to be added together – ones to ones, tens to tens, and hundreds to hundreds.	1. (a) 5 (b) 50 (c) 500 2 ones + 3 ones = 5 ones 2 + 3 = 5 (1)(1)(1)(1)(1) 2 + 3 5 2 tens + 3 tens = 5 tens 20 + 30 = 50 (10)(10)(10)(10)(10) 2 0 + 3 0 5 0 2 hundreds + 3 hundreds = 5 hundreds 200 + 300 = 500 (100)(100)(100)(100)(100) 2 0 0 + 3 0 0 5 0 0
Use the addition algorithm	
Write the expression **356 + 31** and show the two numbers on a place-value chart with place-value discs, one above the other. Then write the problem vertically on the board, aligning the digits. Draw a vertical line between the digits to emphasize that the ones, tens, and hundreds of both numbers are lined up. Tell students that we are going to add starting at the smallest place value this time. Ask them which is the smallest place value. It is the ones place. Ask them to add the ones.	356 + 31 3 ¦5 ¦6 + ¦3 ¦1 ¦ ¦**7** Add the ones. ↓

Ask them to add the tens, and then find the hundreds, as you combine the discs and write the answers.	3 5 6 + 3 1 = **8** 7 — Add the tens. ↓ 3 5 6 + 3 1 = **3** 8 7 — Write the hundreds.
Add	**Text p. 29**
Write the problems on the board horizontally and ask students to copy, aligning the digits, or have students copy from the textbook. Emphasize that the digits for each place value should be lined up on top of each other, tens of one number above the tens of the other number, and ones of the first number above the ones of the second number.	2. 25 + 32 = **57** 4. 251 + 34 = **285**
Assessment	**Text p. 29**
Task 3: More capable students may be able to solve some of these mentally. Those they cannot they should rewrite vertically and use the addition algorithm. You can let less capable students use number discs and a place-value chart.	3. (a) 61 + 8 = **69** (b) 75 + 4 = **79** (c) 34 + 24 = **58** (d) 19 + 50 = **69** (e) 60 + 34 = **94** (f) 70 + 29 = **99**
Write some additional expressions on the board and have students copy them and solve them. This may be mostly an exercise in writing the problems vertically for some students, since more capable students will be able to solve these mentally, as they have been doing with 2-digit numbers earlier.	104 + 32 (136) 730 + 60 (790) 325 + 72 (397) 239 + 50 (289) 534 + 15 (549) 964 + 22 (986)
Practice	WB Exercise 11, p. 30, problems 1-2 WB Exercise 12, p. 32, problem 1

Exercise 11

1. (a) 8 ones; 8
   (b) 8 tens; 80
   (c) 8 hundreds; 800

2. (a) 7, 70, 700 (b) 10, 100, 1000

30

Exercise 12

1. 598 396 787

   189 856 495

   655 789 971

   MEET ME AT THE CORNER

32

2.2b Add a 3-digit Number

Objectives

- Add a 3-digit number to a number within 1000 without renaming, using the standard algorithm.
- Add numbers within 1000, no renaming.

Note

Show the process using place-value discs and relate the steps to the written problem by putting the discs for both numbers on the chart, then combining the ones as you add ones, the tens as you add tens, and the hundreds as you add hundreds.

You may want to let students follow along with some of the problems using their own discs. This may not be necessary for this lesson, since there is no renaming.

Discuss addition of 3-digit numbers	
Write the expression **642 + 146.** Show the two numbers on a place-value chart with place-value discs, one above the other. Then write the problem vertically, aligning the digits. Ask students to add the ones, then the tens, then the hundreds, as you write the answers in the correct place.	642 + 146 $\begin{array}{r} 6\ 4\ 2 \\ +\ 1\ 4\ 6 \\ \hline \mathbf{8} \end{array}$ Add the ones. ↓ $\begin{array}{r} 6\ 4\ 2 \\ +\ 1\ 4\ 6 \\ \hline \mathbf{8}\ 8 \end{array}$ Add the tens. ↓ $\begin{array}{r} 6\ 4\ 2 \\ +\ 1\ 4\ 6 \\ \hline \mathbf{7}\ 8\ 8 \end{array}$ Add the hundreds.
Add	**Text pp. 28, 30**
Task 5: Write the problem on the board and guide students in solving it. Task 7: We need to find the total, which is the total number of cartons he sold both days. We add.	236 + 362 = **598** 5. 245 + 142 = **387** 7. 124 + 65 = **189**
Assessment	**Text p. 30**
Task 6: Students should rewrite any expressions they cannot solve mentally and solve using the addition algorithm. You can have them use the textbook or write the problems on the board to be sure they are working on the correct problems.	6. (a) 104 + 30 = **134** (b) 230 + 60 = **290** (c) 125 + 72 = **197** (d) 539 + 50 = **589** (e) 442 + 134 = **576** (f) 342 + 253 = **595**

Write some additional expressions on the board for students to solve. They may need to practice rewriting the problems vertically and aligning the digits correctly, even if they can solve them mentally.	412 + 157 (569) 730 + 269 (999) 125 + 724 (849) 232 + 506 (738)
Write an addition word problem, such as the following, on the board and have your students solve it. ⇒ Harriet sold one of her macaws for $432 and another for $327. How much money did she make from selling these two birds?	$432 + $327 = $759 She made $759.
Enrichment	Mental Math 12
Practice	WB Exercise 12, pp. 33-34, Problems 2-5

Game

This game is meant to prepare students for addition with renaming which they will be learning later in this unit. They can play it now or later, as time permits, before Part 4 of this unit.

Material: Number cards 0-9, 4 sets for each group, shuffled and placed face-down in the middle. Place-value discs and charts for each student.

Procedure: Divide students into groups. Announce a target number, such as 500 or 900. Players take turns drawing a card. For each number they draw, they place the same number of ones on their chart. Whenever there are 10 in the ones or tens column, they need to be traded in for a disc of the next higher place value. The first student who reaches the target number (e.g., has 5 hundred-discs in the hundreds column) wins. If several students have reached the target number in the round, the one closest (with fewest discs in the other columns) wins.

Exercise 12

2. 849 667 835

789 798 366

488 569 987

Sentosa Island

33

3. 410 + 56 = 466
466

4. 125 + 63 = 188
188

5. 242 + 304 = 546
546

34

2.3 Subtraction Without Renaming

Objectives

- Subtract a 2-digit number from a number within 1000, no renaming.
- Subtract a 3-digit number from a number within 1000, no renaming.
- Add and subtract numbers within 1000, no renaming.
- Solve word problems involving addition and subtraction.

Material

- Place value discs and charts for students and board
- Number cards 0-10, 4 sets per group
- Mental Math 13-14 (appendix)

Prerequisites

Students should have a good knowledge of place value and of the basic subtraction facts within 10.

Notes

Up to this point, students have been subtracting 2-digit numbers in a horizontal format using a variety of strategies, including subtracting the tens first and then the ones. Here, the standard algorithm for subtraction is introduced where the problem is worked in a vertical format. Ones are subtracted first, then tens, and so on.

In this part students will be subtracting numbers using the standard algorithm where renaming does not occur. In Part 5 of this unit, they will be subtracting numbers where renaming does occur. Working with place-value discs lets students become comfortable with adding and subtracting larger numbers, and later with renaming.

As students become more proficient with subtracting 3-digit numbers using the vertical format, some may be able to solve the problems in a horizontal format, particularly if they have a good sense of place value and can align the digits mentally. However, they should always be able to use the standard algorithm if the problem is too difficult to do mentally, such as when renaming occurs over several place values. The standard algorithm is a single approach that will work in all circumstances; mental math strategies vary depending on the numbers involved.

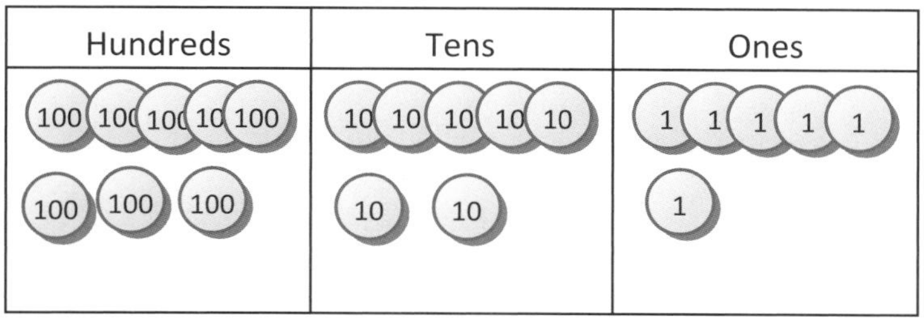

Hundreds	Tens	Ones
100 100 100 100 100 100 100 100	10 10 10 10 10 10 10	1 1 1 1 1 1

```
  8 7 6
– 5 3 2
```

Subtract the ones.

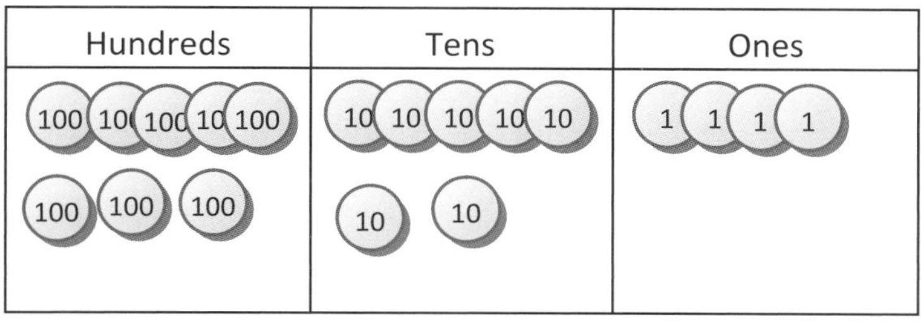

Hundreds	Tens	Ones
100 100 100 100 100 100 100 100	10 10 10 10 10 10 10	1 1 1 1

```
  8 7 6
– 5 3 2
      4
```

Subtract the tens.

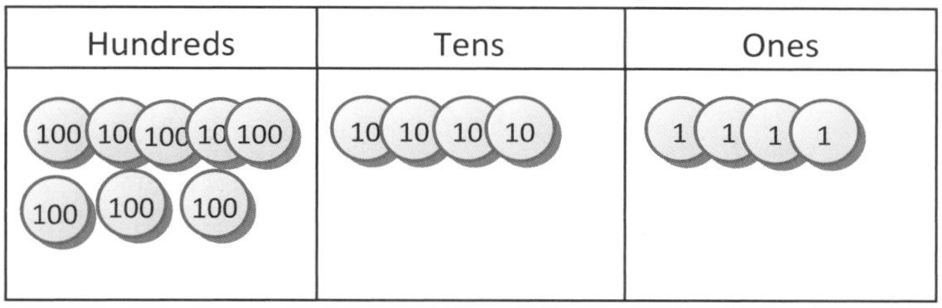

Hundreds	Tens	Ones
100 100 100 100 100 100 100 100	10 10 10 10	1 1 1 1

```
  8 7 6
– 5 3 2
    4 4
```

Subtract the hundreds.

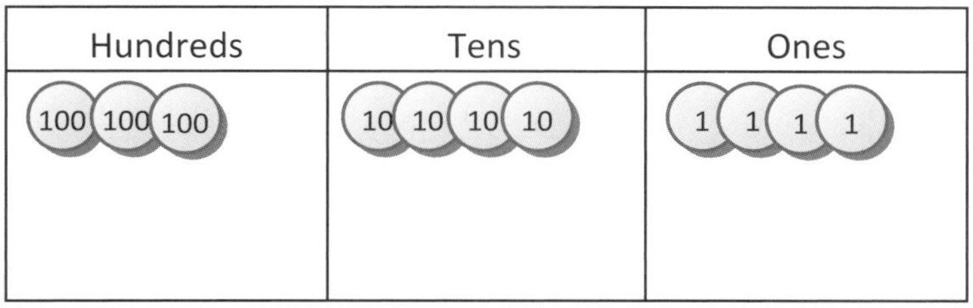

Hundreds	Tens	Ones
100 100 100	10 10 10 10	1 1 1 1

```
  8 7 6
– 5 3 2
  3 4 4
```

2.3a Subtract a 2-digit Number

Objectives

- Subtract a 2-digit number from a number within 1000 without renaming, using the standard algorithm.

Note

Students with a good understanding of place value will be able to do the problems in this lesson mentally. They should still practice writing at least some of them vertically, and aligning the digits correctly. They will better understand the usefulness of the algorithm when renaming occurs.

Subtract ones, tens, and hundreds	**Text p. 32**
If necessary, illustrate the problems with place-value discs on the board. Allow students to follow along with their own discs after the initial example.	1. (a) 4 (b) 40 (c) 400
Rewrite the problems vertically. Remind students that if we write the numbers one above the other, making sure the digits for each place-value in the first number line up with the digits for the same place-value in the second number, it is easy to see which digits need to be subtracted from which – ones from ones, tens from tens, or hundreds from hundreds.	1 1 1 1 1 1 1 7 ones – 3 ones = 4 ones 7 – 3 = 4 $\begin{array}{r} 7 \\ -\ 3 \\ \hline 4 \end{array}$ 10 10 10 10 10 10 10 7 tens – 3 tens = 4 tens 70 – 30 = 40 $\begin{array}{r} 70 \\ -\ 30 \\ \hline 40 \end{array}$ 100 100 100 100 100 100 100 7 hundreds – 3 hundreds = 4 hundreds 700 – 300 = 400 $\begin{array}{r} 700 \\ -\ 300 \\ \hline 400 \end{array}$
Use the subtraction algorithm	
Write the expression **698 – 45** and show 698 on a place-value chart with place-value discs. Then write the problem vertically, aligning the digits. You can draw vertical lines between the place values to emphasize that the ones, tens, and hundreds of both numbers are lined up. Ask students to subtract the ones. Remove 5 ones from the chart and write the ones for the answer.	698 – 45 $\begin{array}{r} 698 \\ -\ 45 \\ \hline \mathbf{3} \end{array}$ Subtract the ones. ↓

Repeat with the tens, and then write the hundreds.	6 ¦ 9 ¦ 8 − ¦ 4 ¦ 5 ¦ **5** ¦ 3 Subtract the tens. ↓ 6 ¦ 9 ¦ 8 − ¦ 4 ¦ 5 **6** ¦ 5 ¦ 3 Write the hundreds.
Subtract	**Text p. 32**
Write the expressions on the board horizontally and ask students to copy, aligning the digits. They could also copy from the textbook. Guide students in subtracting these numbers, starting with the ones.	2. 36 – 12 = **24** 4. 239 – 25 = **214**
Assessment	**Text p. 32**
You may want to write the problems horizontally on the board to make sure students are working on the correct ones. Students should copy, putting the numbers in a vertical format.	3. (a) 78 – 4 = **74** (b) 78 – 40 = **38** (c) 65 – 5 = **60** (d) 65 – 50 = **15** (e) 59 – 37 = **22** (f) 48 – 38 = **10**
Write some additional expressions on the board and have students copy them, rewriting them vertically, and solve them using the subtraction algorithm. Some students can solve these mentally but may need practice rewriting the problems and aligning digits.	164 – 32 (132) 780 – 60 (720) 325 – 12 (313) 289 – 50 (239) 537 – 15 (522) 964 – 22 (942)
Practice	WB Exercise 13, p. 35, problems 1-2 WB Exercise 14, p. 37, problem 1

Exercise 13

1. (a) 6 ones; 6
   (b) 6 tens; 60
   (c) 6 hundreds; 600

2. (a) 4, 40, 400   (b) 9, 90, 900

35

Exercise 14

1. 657 713 908

   120 326 549

   834 245 400

37

2.3b Subtract a 3-digit Number

Objectives

- Subtract a 3-digit number from a number within 1000 without renaming, using the standard algorithm.

Note

Show the process using place-value discs and relate the steps to the written problem by putting the discs for the total on the chart, then removing ones as you subtract ones, tens as you subtract tens, and hundreds as you subtract hundreds.

You may want to let students follow along with some of the problems using their own discs. For those that need it, they should work on problems using discs individually after your example.

<table>
<tr><th>Discuss subtraction of 3-digit numbers</th><th></th></tr>
<tr><td>Write the expression 948 – 345. Show 948 with place-value discs. Then write the problem vertically, aligning the digits. Ask students to subtract the ones, then the tens, then the hundreds, as you remove the discs and write the answers in the correct place.</td><td>948 – 345

  9 4 8
– 3 4 5
     3 Subtract the ones.
↓
  9 4 8
– 3 4 5
   0 3 Subtract the tens.
↓
  9 4 8
– 3 4 5
  6 0 3 Subtract the hundreds.</td></tr>
<tr><th>Subtract</th><th>Text pp. 31, 33</th></tr>
<tr><td>Discuss the problem on p. 31 of the text. You can write the numbers on the board and go through the subtraction steps.

Task 5: Write the expression on the board and guide students in solving it.

Task 7: We subtract to find a missing part.</td><td>396 – 214 = 182; 182

5. 376 – 152 = 224

7. 287 – 52 = 235; 235</td></tr>
<tr><th>Assessment</th><th>Text p. 33</th></tr>
<tr><td>You may want to write the problems on the board.

Some students may be able to solve them mentally.</td><td>6. (a) 486 – 80 = 406
(b) 178 – 100 = 78
(c) 597 – 85 = 512
(d) 269 – 62 = 207
(e) 365 – 145 = 220
(f) 486 – 160 = 326</td></tr>
</table>

Write some additional expressions on the board for students to solve using the subtraction algorithm.	479 – 157 (322) 760 – 230 (530) 725 – 404 (321) 999 – 123 (876)
Write a word problem, such as the following, on the board and have students solve it. ⇒ Harriet sold one of her macaws for $425 and then bought another bird for $322. How much more did the bird she sold cost than the one she bought?	$425 – $322 = $103 It cost $103 more.
Enrichment	Mental Math 13
Practice	WB Exercise 14, pp. 38-39, Problems 2-5

Game

Students will be learning subtraction with renaming later in this unit, and this game is meant to prepare students for subtraction with renaming. They can play it now or later, as time permits, before Part 5 of this unit.

Material: Number cards 0-9, 4 sets for each group, shuffled and placed face-down in the middle. Place-value discs and charts for each student.

Procedure: Divide students into groups. Announce a starting number such as 555. Each student should place the correct number of discs on the chart to represent this number. Players take turns drawing a card. For each number they draw, they remove the same number of ones from their place-value chart as the number on the card, trading a hundred for 10 tens and a ten for 10 ones when needed. The first player who does not have enough ones to remove (the number drawn is higher than the number of ones left on the chart) wins.

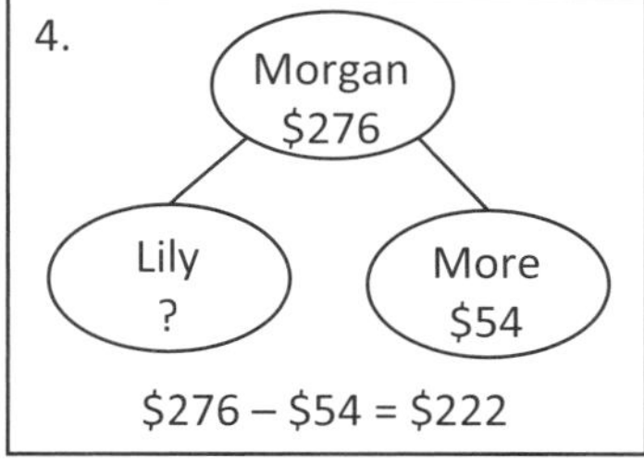

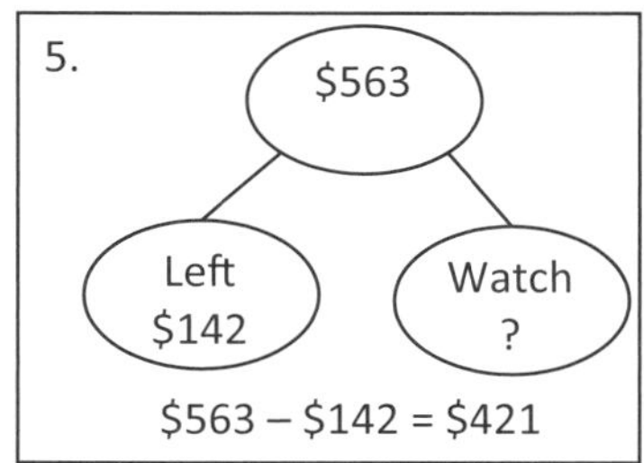

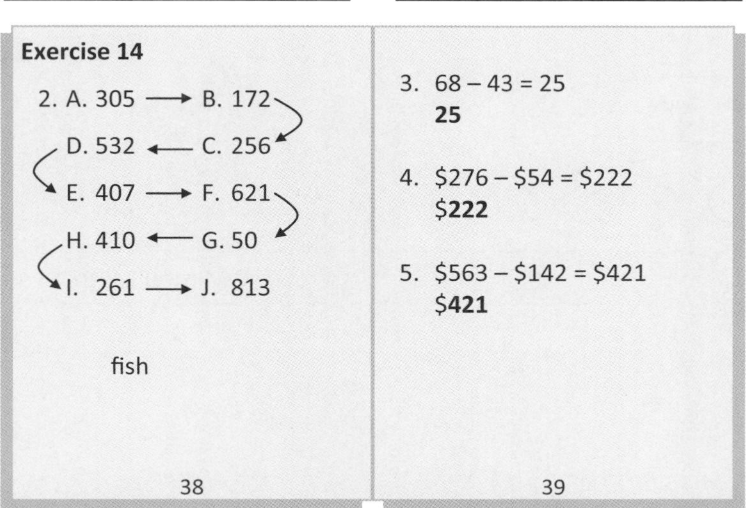
Exercise 14

2. A. 305 → B. 172 → C. 256 → D. 532 → E. 407 → F. 621 → G. 50 → H. 410 → I. 261 → J. 813

fish

38

3. 68 – 43 = 25
25

4. $276 – $54 = $222
$222

5. $563 – $142 = $421
$421

39

2.3c Practice

Objectives

- Add and subtract within 1000, no renaming.
- Solve word problems involving addition and subtraction.

Note

You can do the problems in the practice as a class, or have students work on them individually and then share their solutions for the word problems. You can do the first two word problems as a class and have students work on the rest on their own. Included below are some suggested questions to help students with the first two problems. If you do not have time for a separate lesson, you can save the problems and use them occasionally during the units on measurement or multiplication and division for more continuous review.

<table>
<tr><th>Practice</th><th>Text p. 35, Practice 2B</th></tr>
<tr><td>Problem 6: Ask students: What do we need to find? (How many buns she had at first.) What information are we given? (How many buns she sold, and how many she had left.) Are we given a whole? Did she start with more buns than she gave away? We need to find the total she started with. Draw a number bond. We need to add the number of buns she sold and the number she did not sell in order to find how many buns she had at first.

Problem 7: What do we need to find? (How many more English books there are than Spanish books.) Which are there more of, English or Spanish books? (English) So there are fewer Spanish books, and we want to find how many fewer. Draw a number bond. We can draw a bigger oval for English books, because there are more of them. To find out how many more, do we add or subtract? We subtract. (If students have trouble, rephrase the problem with 5 English books and 2 Spanish books, and then return to the original problem.)
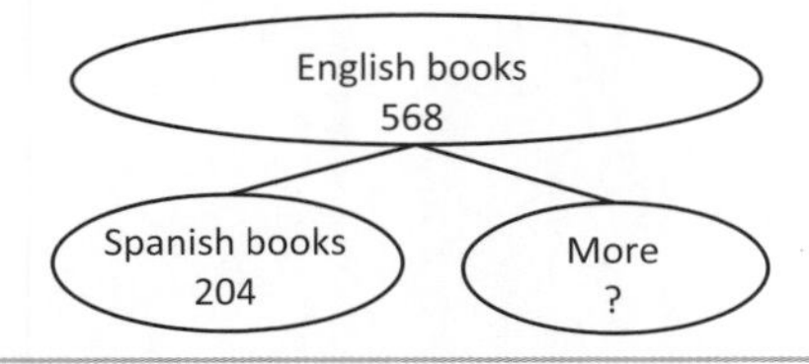
</td><td>1. (a) 359 (b) 168 (c) 599
2. (a) 862 (b) 622 (c) 441
3. (a) 193 (b) 567 (c) 597
4. (a) 528 (b) 294 (c) 224
5. (a) 488 (b) 502 (c) 607
6. 245 + 54 = 299
She had 299 buns at first.
7. 568 – 204 = 364
There are 364 more English books.
8. 439 – 326 = 113
She had 113 eggs left.
9. 768 – 532 = 236
236 were children.
10. (a) 104 + 125 = 229
229 children took part.
(b) 125 – 104 = 21
There were 21 more girls than boys.</td></tr>
<tr><td>Enrichment</td><td>Mental Math 14</td></tr>
</table>

Group Game (Review of addition and subtraction facts through 20.)

Material: Number cards 1-10, 4 sets for each group.

Procedure: Divide students into groups. One student is the dealer and shuffles the cards and deals 5 to each player. The remaining cards are placed face down in the middle. The players take turns turning over 2 cards from the middle and adding the numbers together. If he or she has two cards with the same sum, then that player lays those cards down. If not, he or she says "Nope" (or some other distinctive word) and then any other player can lay down two cards with the same sum. The one that is fastest and lays cards down first can leave them down; the rest who also have two cards with the same sum have to return theirs to their hand. The first player to get rid of all of his or her cards wins.

Variation: Three cards are turned over, instead of two. The two highest cards are added together. This sum is the total. The remaining card is a part. If the player has a card that can be the other part, he or she lays that one down.

2.4 Addition With Renaming

Objectives

- Add ones or tens to a 3-digit number using mental math strategies.
- Add within 1000.
- Add three numbers.
- Solve word problems.

Material

- Place-value discs and charts
- Number cube 4-9, 1 per group
- Number cards 1-9, 4 sets per group
- Mental Math 15 (appendix)

Prerequisites

Students should have a good knowledge of place value and of the basic addition facts within 20. For the lesson on mental math, they should be familiar with the "make a ten" strategy as reviewed in lesson 1.1d.

Notes

In *Primary Mathematics* 1B students learned to add a 1-digit number to a 2-digit number mentally when renaming occurs. This will be reviewed in the first lesson for this part and extended to adding ones or tens to a 3-digit number. If too many students struggle with mental math, save this lesson for later, do the first question on p. 37 and the first page of Exercise 15 with the next lesson, allowing students to rewrite the problems vertically and use the standard addition algorithm. Introduce the concepts in the first lesson in this guide as time permits and as you feel the students might be ready for it as you proceed.

In this part students will add numbers within 1000 where renaming occurs. They will first add numbers where renaming occurs only in the ones, then where renaming occurs only in the tens, and finally where renaming occurs in both the tens and ones.

As you guide students through the algorithm for addition, show the steps initially with place-value discs and a place-value chart, going through the process step-by-step. These steps are shown on the next page for an addition problem that involves renaming in both the ones and the tens places. For less capable students you can first show the steps with the discs, but without the written problem, and then show the steps again while relating the written problem to each step. In the written problem we are recording what we do with the discs. Students should understand how the process works concretely and in terms of place-value, and not simply memorize the steps in a written problem. Let them use the place-value discs as needed. Eventually, they should be able to add the numbers without the chart and discs.

When discussing the steps *always* use the place-value name with the digit. In the example in the next page, we are adding 6 *tens* and 8 *tens*, *not* 6 and 8.

Addition with larger numbers can be time-consuming for students at first. Take the time needed for them to thoroughly understand the process. Then, as you continue with the next units, continue to give them one or two problems to practice each day. On the other hand, if students are understanding the algorithm easily, spend more time with word problems from some of the supplementary books.

If a student is good at mental math, he or she may be able to come up with mental techniques for adding 3-digit numbers with renaming, such as starting with the hundreds and looking ahead to see how the tens will affect the hundreds before writing down the answer. Some specific strategies will be taught in *Primary Mathematics* 2B, and additional strategies in *Primary Mathematics* 3. Allow students to develop flexibility in working with numbers through experimenting with different ways of renaming, and to decide which method is best to use when working independently to arrive at the correct answer. However, all students should be thoroughly familiar with the standard addition algorithm, since it can be used in all cases, whereas mental math strategies vary depending on the problem.

The advantage to starting with the ones place in the standard algorithm is that it avoids retracing steps. Once the ones are found, they will not change. Then when the tens are found next, they too will not change. In working from left to right instead, and starting with the hundreds, adding the ones can change both the hundreds and the tens.

Hundreds	Tens	Ones
100 100 100	10 10 10 10 10 10	1 1 1 1 1 1 1 1
100 100 100 100	10 10 10 10 10 10 10 10	1 1 1 1 1

$$\begin{array}{r} 3\;6\;8 \\ +\;4\;8\;5 \\ \hline \end{array}$$

Add the ones:
8 ones + 5 ones = 13 ones
Rename:
13 ones = 1 ten 3 ones
Write the 1 ten above the tens place and the total ones under the line in the ones place.

Hundreds	Tens	Ones
	10	1 1 1 1 1 1 1 1 1 1
100 100 100	10 10 10 10 10 10	1 1 1
100 100 100 100	10 10 10 10 10 10 10 10	

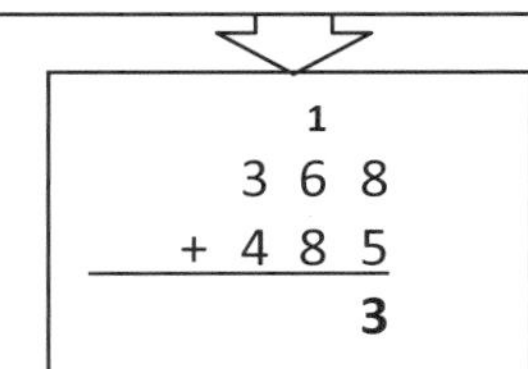

Add the tens:
6 tens + 8 tens + 1 ten = 15 tens
Rename:
15 tens = 1 hundred 5 tens
Write the 1 hundred above the hundreds place and the total tens under the line in the tens place.

Hundreds	Tens	Ones
100	10 10 10 10 10 10 10 10 10 10	1 1 1
100 100 100		
100 100 100 100	10 10 10 10 10	

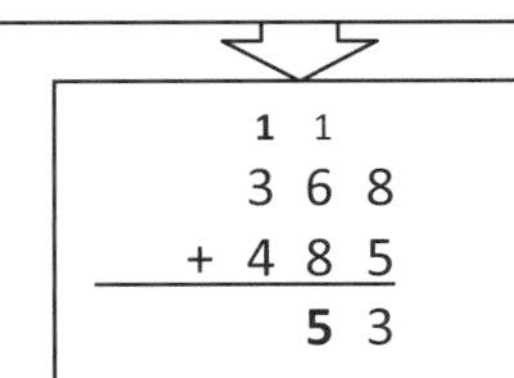

Add the hundreds:
3 hundreds + 4 hundreds + 1 hundred = 8 hundreds
Write the total hundreds under the line in the hundreds place.

Hundreds	Tens	Ones
100 100 100 100 100 100 100 100	10 10 10 10 10	1 1 1

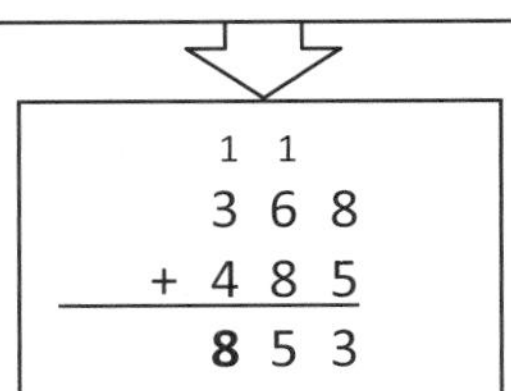

2.4a Add Ones or Tens to a 3-Digit Number

Objectives

Add ones or tens to a 3-digit number using mental math strategies.

Note

In this lesson, students will add ones or tens to a 3-digit number. The mental math strategies are an extension of adding ones mentally, even when there is renaming, as in 4 + 9, which students have already learned. Students can use either a "make a ten" strategy or addition facts. Those students who have difficulty memorizing addition facts will be able to use the "make a ten" strategy in order to make accurate calculations rather than relying on faulty memory. A student who does not memorize facts easily can still be exceptional in math ability.

Mental math can be challenging for other students. If they struggle with mental math, they can always fall back on the standard algorithm, which they will learn in the next lesson. Students can also solve the problems by counting up in the correct place and keeping track with fingers. For example, 462 + 70: Count up by tens from 462: 47 (tens), 48, 49, 50, 51, 52, 53 → 532.

You can draw number bonds to illustrate the thinking process. However, the purpose of this lesson is for students to learn mental math strategies and use them, not to substitute an alternate paper-and-pencil method for addition that can only be used in certain circumstances. If students need to do these problems using paper and pencil, then they need to go back to the earlier lessons in *Primary Mathematics* 1 where this type of number sense is developed, and in the meantime they should simply understand and use the standard algorithm rather than having to decide which paper and pencil strategy to use when.

Illustrate any of the problems in this lesson with place-value discs, as needed.

Add ones	
Write the problem **8 + 7 = 10 + ___ = ____** on the board. Ask students fill in the blank to find the answer by making a ten and then find the answer. Tell students that we can also find the answer by simply remembering the math fact 8 + 7 = 15.	8 + 7 = 10 + ____ = ____ 8 + 7 = 10 + 5 = 15 2 5
Write the expression **34 + 2** on the board. Ask students if adding the ones will result in an answer greater than 10. It does not. So we can simply add the ones to find the answer.	34 + 2 34 + 2 = 36
Write the expression **38 + 7**. Ask students if adding the ones will result in an answer greater than 10. It does. Ask students what the next ten is after 38. How much do we need to add to 38 to get to 40? If we take the 2 from the 7, how much is left? (5) So the answer is 40 and 5, or 45. Tell students that we can also add by splitting 30 into 30 and 8 if we remember that 8 + 7 = 15. Then we can simply add 15 to 30.	38 + 7 38 + 7 = 40 + 5 = 45 2 5 38 + 7 = 30 + 15 = 45 30 8
Write the expression **638 + 7**. Ask students if adding the ones will give more than ten ones. It will; there will be another ten. Ask if there will be another hundred. There will not. We can add 638 + 7 in the same way as we would add 38 and 7, but include the hundreds in the answer.	638 + 7 638 + 7 = 640 + 5 = 645 2 5

Add tens	
Write the expression **80 + 70** and ask students to find the answer. Since 8 ones + 7 ones = 15 ones, then 8 tens + 7 tens = 15 tens, or 150. We can also make the next hundred, by adding 20 to 80, leaving 50.	80 + 70 80 + 70 = 100 + 50 = 150 /\\ 20 50
Ask students how many total tens are in 300 if we only used ten-discs for the number. Since each hundred is 10 tens, 3 hundreds is 30 tens.	300 = 30 tens
Ask students how many tens are in 380. There are 30 tens and 8 more tens, or 38 tens.	380 = 38 tens
Write the expression **380 + 70**. Point out that we are adding 38 tens and 7 tens. Since 38 ones + 7 ones = 45 ones, 38 tens + 7 tens = 45 tens, or 450.	380 + 70 380 + 70 = 400 + 50 = 450 /\\ 20 50
Write the expression **384 + 70**. Point out that we are simply adding tens, and the ones do not change. We just have to add 38 tens and 7 tens, write that down, and add the 4 in the ones place.	384 + 70 384 + 70 = 454 /\\ /\\ 4 380 20 50
Assessment	**Text p. 37**
Task 1: Have students do these in the order (a), (d), (b), (e), (c), (f). You can write them in that order on the board.	1. (a) 4 + 9 = **13** (d) 40 + 90 = **130** (b) 60 + 9 = **69** (e) 600 + 9 = **609** (c) 64 + 9 = **73** (f) 640 + 90 = **730**
Reinforcement	Mental Math 15
Practice	WB Exercise 15, p. 40, Problems 1-2

Exercise 15

1. (a) 13, 33, 533 (b) 10, 50, 250
   (c) 13, 130, 530 (d) 10, 100, 400
   (e) 17, 77, 277 (f) 10, 90, 390
2. (a) 71 (b) 81
   (c) 145 (d) 640
   (e) 150 (f) 100
   (g) 310 (h) 500

40

2.4b Rename Ones

Objectives

- Add within 1000 by renaming ones.

Vocabulary

- Rename

Note

Show the steps with a place-value charts and discs, and then have students to use place-value discs of their own so that they understand the procedure concretely. Students should then attempt the practice problems without discs, using discs only if they need to. You may have to spend extra time with some students or extend the lesson another day until they are competent in doing the problems without discs. If renaming the ones is thoroughly understood, the rest of the lessons in this part will be easier. More capable students will not need the discs for long.

<table>
<tr><th>Discuss addition with renaming in the ones</th><th></th></tr>
<tr><td>
Write the expression 36 + 8. Students will probably be able to solve this mentally if they have done Primary Mathematics 1B or the previous lesson. Then rewrite it vertically. Show the numbers with place-value discs and ask students to add the ones. Since we cannot write 14 ones down as part of the answer, we have to make a group of ten ones and replace it with a ten. Then we can add the tens.
Show how we record the steps with the written problem. We have 14 ones, which is a ten and a one. We are renaming 14 ones as 1 ten 4 ones. Since we still have tens to add, we write the renamed 10 as a 1 above the tens. By writing it above the tens, we show that it is a ten, not a one. Then we write the 4 ones below the line, since this is the final amount of ones. Only final amounts for each place go below the line.
Then, we add the tens (the 1 ten that was renamed, and the 3 tens we started with) and write the answer below the line in the tens place.
</td><td>
36 + 8
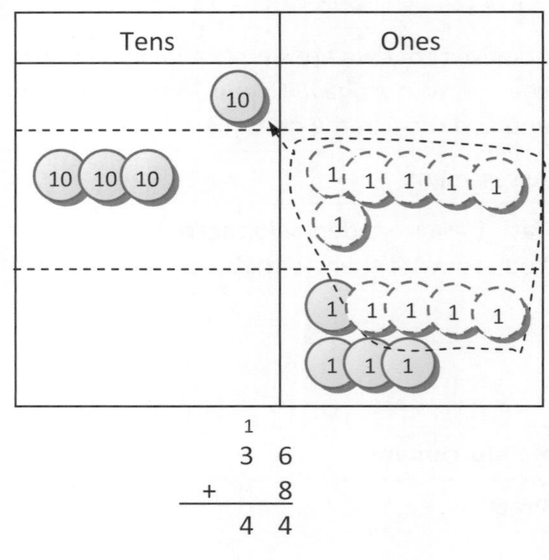

</td></tr>
<tr><td>
Write the expression 36 + 28 and show the two numbers with place-value discs or have students show them with their discs. Ask them to first add the ones and rename them. Show how the process is recorded with the written problem. (This problem is on p. 36 of the textbook.) This is similar to the previous problem, except that now both numbers have tens.
</td><td>
36 + 28
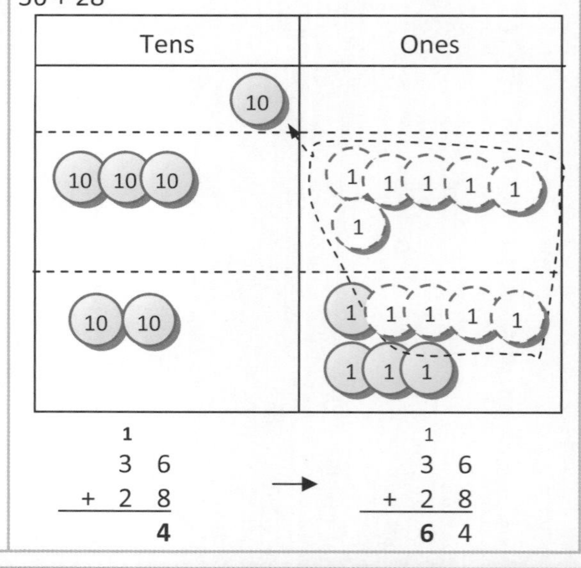

</td></tr>
</table>

Repeat with **536 + 28**. We add the ones and tens in the same way, but now have to write the hundreds.	$\begin{array}{r} {}^{1} \\ 5\ 3\ 6 \\ +\quad 2\ 8 \\ \hline 5\ 6\ 4 \end{array}$
Finally, repeat with **536 + 328**. We add the ones and tens in the same way, and then add the hundreds.	$\begin{array}{r} {}^{1} \\ 5\ 3\ 6 \\ +\ 3\ 2\ 8 \\ \hline 8\ 6\ 4 \end{array}$
Add	**Text pp. 37-38**
Have students use place-value discs. Guide them through the steps.	3. 318 + 43 = **361** 5. 267 + 123 = **390**
Assessment	**Text pp. 37-38**
Have students do some or all of these problems. You can write selected ones from each set on the board and add more if time permits.	2. (a) 35 + 7 = **42** (b) 75 + 5 = **80** (c) 48 + 38 = **86** (d) 54 + 29 = **83** (e) 57 + 13 = **70** (f) 69 + 31 = **100** 4. (a) 315 + 8 = **323** (b) 224 + 7 = **231** (c) 527 + 45 = **572** (d) 608 + 48 = **656** (e) 734 + 36 = **770** (f) 321 + 69 = **390** 6. (a) 127 + 365 = **492** (b) 452 + 219 = **671** (c) 639 + 124 = **763** (d) 745 + 136 = **881** (e) 506 + 104 = **610** (f) 828 + 162 = **990**
Practice	WB Exercise 15, p. 41, problem 3 WB Exercise 16, p. 42, problem 1

Exercise 15

3. A 81 B 92 D 93

H 72 I 82 P 95

R 84 T 70 Y 80

HAPPY BIRTHDAY

41

Exercise 16

1. 981 373 471

793 872 376

750 675 890

42

2.4c Rename Tens

Objectives

- Add within 1000 by renaming tens.

<table>
<tr><th>Discuss renaming in the tens</th><th></th></tr>
<tr><td>

Write the expression **73 + 82** and show the two numbers with place-value discs.

Ask students to add the ones and give you the answer. Ask them where you should write the answer (under the line, in the ones place).

Ask students to add the tens, 7 tens + 8 tens. The answer is not 15, it is 15 tens, or 1 hundred 5 tens. Replace the 15 tens with 1 hundred and 5 tens. Tell them we are renaming 15 tens as 1 hundred 5 tens. Ask them where you should write the two digits. There are no hundreds still to add, so they both can go under the line.

</td><td>

73 + 82

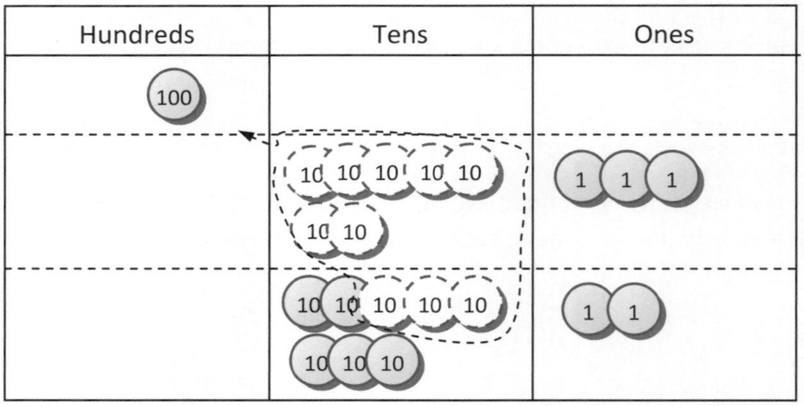
Hundreds
Tens
Ones
100
10 10 10 10 10
10 10
1 1 1
10 10 10 10 10
10 10 10
1 1

$\begin{array}{r} 7\ 3 \\ +\ 8\ 2 \\ \hline \end{array} \rightarrow \begin{array}{r} 7\ 3 \\ +\ 8\ 2 \\ \hline \mathbf{5} \end{array} \rightarrow \begin{array}{r} 7\ 3 \\ +\ 8\ 2 \\ \hline \mathbf{1}\ \mathbf{5}\ 5 \end{array}$

</td></tr>
<tr><td>

Repeat with **373 + 82**. This time, when we add the tens to get 1 hundred 5 tens, there still hundreds to add, so the 1 hundred they get from adding tens is not the final number of hundreds. We record this hundred by writing the 1 above the hundreds place. Then we can add that hundreds to the rest of the hundreds to get the final number of hundreds, which we write below the line.

</td><td>

373 + 82

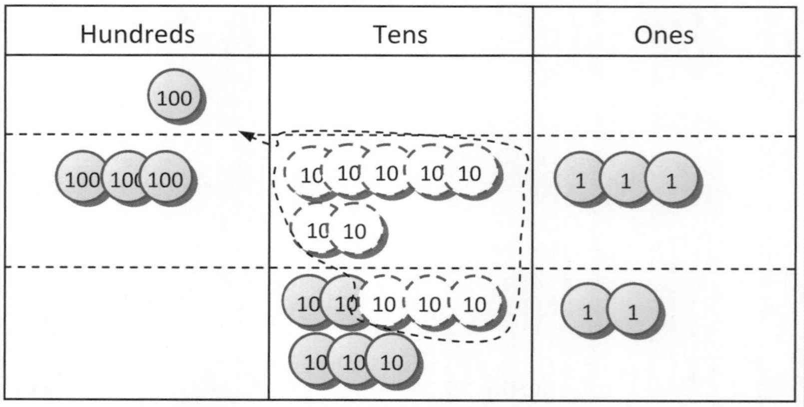
Hundreds
Tens
Ones
100
100 100 100
10 10 10 10 10
10 10
1 1 1
10 10 10 10 10
10 10 10
1 1

$\begin{array}{r} 3\ 7\ 3 \\ +\ \ \ 8\ 2 \\ \hline \mathbf{5} \end{array} \rightarrow \begin{array}{r} \mathbf{1}\ \ \ \ \ \ \\ 3\ 7\ 3 \\ +\ \ \ 8\ 2 \\ \hline \mathbf{5}\ 5 \end{array} \rightarrow \begin{array}{r} 1\ \ \ \ \ \ \\ 3\ 7\ 3 \\ +\ \ \ 8\ 2 \\ \hline \mathbf{4}\ 5\ 5 \end{array}$

</td></tr>
<tr><td>

Repeat with **373 + 482**. Emphasize that we first fill in the ones place under the line with the total ones, then the tens place with the total tens, then the hundreds place with the total hundreds.

</td><td>

373 + 482

$\begin{array}{r} 3\ 7\ 3 \\ +\ 4\ 8\ 2 \\ \hline \mathbf{5} \end{array} \rightarrow \begin{array}{r} \mathbf{1}\ \ \ \ \ \ \\ 3\ 7\ 3 \\ +\ 4\ 8\ 2 \\ \hline \mathbf{5}\ 5 \end{array} \rightarrow \begin{array}{r} 1\ \ \ \ \ \ \\ 3\ 7\ 3 \\ +\ 4\ 8\ 2 \\ \hline \mathbf{8}\ 5\ 5 \end{array}$

</td></tr>
</table>

Add	Text pp. 38-39
Have students use place-value discs, and guide them through the steps.	7. 563 + 56 = **619** 9. 382 + 145 = **527**
Assessment	**Text pp. 38-39**
Have students do some or all of these. You can write the ones you want them to do on the board.	8. (a) 292 + 60 = **352** (b) 574 + 70 = **644** (c) 385 + 63 = **448** (d) 630 + 94 = **724** (e) 420 + 80 = **500** (f) 279 + 30 = **309** 10. (a) 454 + 163 = **617** (b) 670 + 156 = **826** (c) 257 + 351 = **608** (d) 588 + 220 = **808** (e) 363 + 255 = **618** (f) 790 + 139 = **929**
Practice	WB Exercise 16, p. 43, problem 3

Group Game: Roll 500

Material: Number cube 4-9, 1 per group. Paper with three columns, one each for ones, tens, and hundreds.

Procedure: Each player rolls the number cube once and writes the number down in either the ones or the tens column. If it is in the tens column, the player writes a 0 in the ones column. Each rolls the cube a second time, decides whether it is to be tens or ones, and adds it to the previous number. Each player rolls the cube a third time, decides whether the number is to be tens or a ones, and adds it to the previous sum. The play continues until each player has rolled the cube 10 times. The goal is for the final sum to be as close to 500 as possible.

1st roll, 6	60
2nd roll, 5	+ 50
	110
3rd roll, 8	80
	190
4th roll, 9	90
	280
5th roll, 5	50
	330
6th roll, 7	70
	400
7th roll, 8	80
	480
8th roll, 9	9
	489
9th roll, 7	7
	496
10th roll, 6	6
	502

Exercise 16

2. 865 → 435 → 826 ↓

327 ← 787 ← 519

↓

900 → 627 → 318

airplane

43

2.4d Word Problems

Objectives

- Add within 1000 by renaming ones or tens.
- Solve word problems.

Note

Students need to read the word problems carefully. The last lesson was on addition, but not all the word problems will involve addition. For example, problem 6 is a subtraction problem that does not involve renaming. Students need to learn to read the problems carefully and not simply do whatever operation is currently being learned.

Practice addition	**Text pp. 41-42, Practices 2C and 2D**
Write some or all of the problems 1, 2, 4, and 5 from Practice 2C on the board, horizontally. Guide students through several of them, and have them do the rest independently, rewriting them vertically. Repeat with some or all of problems 1 and 2 in Practice 2D.	Practice 2C 1. (a) 26 + 9 = **35** (b) 32 + 8 = **40** (c) 46 + 7 = **53** 2. (a) 35 + 28 = **63** (b) 51 + 29 = **80** (c) 63 + 27 = **90** 4. (a) 27 + 80 = **107** (b) 33 + 82 = **115** (c) 49 + 70 = **119** 5. (a) 53 + 62 = **115** (b) 64 + 65 = **129** (c) 72 + 37 = **109** Practice 2D 1. (a) 264 + 50 = **314** (b) 379 + 60 = **439** (c) 342 + 93 = **435** 2. (a) 407 + 38 = **445** (b) 532 + 48 = **580** (c) 644 + 49 = **693**
Word problems	**Text p. 41, Practice 2C**
Problem 6: Write the problem on the board and discuss the solution. You do not have to use the exact words in any of these problems. For example, you could substitute the following, or anything more relevant to the students. ⇒ Harriet now has so many birds. She counted her parakeets and parrots. She has 92 parakeets and 42 parrots. How many more parakeets does she have than parrots? We are comparing two numbers and finding the difference (how much more one is than the other) so we subtract. If drawing a number bond, the larger number is the whole and the smaller number a part. The difference is the other part.	6. Singapore stamps 92 Malaysian stamps 42	? 92 – 42 = **50** He has 50 more Singapore stamps.

Problem 7: Write the problem or a related one on the board and discuss its solution. The two parts are the number sold and the number left. We need to find the whole, the number she had at first, so we add.	7. 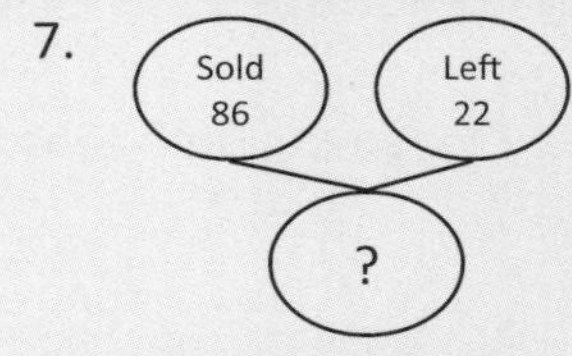 86 + 22 = **108** She had 108 sticks at first.
Assessment	**Text pp. 41-42, Practices 2C and 2D**
You may want to write problems 8 and 9 from Practice 2C and problems 7, 9, and 10 from Practice 2D on the board, or you can have students look at them in their books. By writing them on the board, you can be sure that they are doing the correct problems. You can substitute other words for the problems to make them more interesting. Have students work on the problems individually and then share their solutions. Note that problem 8 on p. 41 and 9 on p. 42 are solved with subtraction (no renaming).	Practice 2C 8. 58 – 42 = **16** He had 16 cards left. 9. 18 + 26 = **44** Her father is 44 years old. Practice 2D 7. 240 + 85 = **325** He bought 325 buttons altogether. 9. 285 – 70 = **215** He sold 215 pizzas. 10. \$125 + \$36 = **\$161** Samy saved \$161.
Practice	WB Exercise 17, pp. 44-45

Exercise 17	
1. 91 56 355 131 523 480 824 403 852 Rabbit	2. 231 + 19 = 250 **250** 3. 285 + 72 = 357 **357** 4. \$162 + \$360 = \$522 **\$522**
44	45

2.4e Rename Ones and Tens

Objectives

- Add within 1000 by renaming both tens and ones.

Discuss renaming in the tens and ones	
Write the expression **269 + 4** on the board. Students should be able to solve this mentally. You can have them show the numbers with place-value discs and discuss the steps in solving this using the standard algorithm.	269 + 4 $\begin{array}{r} 2\ 6\ 9 \\ +\quad\ \ 4 \\ \hline \end{array} \rightarrow \begin{array}{r} ^{1}\quad \\ 2\ 6\ 9 \\ +\quad\ \ 4 \\ \hline \mathbf{3} \end{array} \rightarrow \begin{array}{r} 2\ 6\ 9 \\ +\quad\ \ 4 \\ \hline \mathbf{2}\ \mathbf{7}\ 3 \end{array}$
Write the expression **269 + 34** and show the numbers with place-value discs. Ask students to add the ones first. We need to rename 13 ones as 1 ten and 3 ones. Now, when we add all the tens, including the one from adding the ones, we have 10 tens. We have to rename these as 1 hundred and 0 tens. Then we add the hundreds. Show how these steps are recorded on the written problem.	269 + 34 Hundreds / Tens / Ones Hundreds / Tens / Ones $\begin{array}{r} 2\ 6\ 9 \\ +\ \ 3\ 4 \\ \hline \end{array} \rightarrow \begin{array}{r} ^{1}\quad \\ 2\ 6\ 9 \\ +\ \ 3\ 4 \\ \hline \mathbf{3} \end{array} \rightarrow \begin{array}{r} ^{1\ 1}\quad \\ 2\ 6\ 9 \\ +\ \ 3\ 4 \\ \hline \mathbf{3}\ \mathbf{0}\ 3 \end{array}$
You can repeat with the expression **269 + 434**. The only difference from the previous problem is that there are now - hundreds to add.	$\begin{array}{r} 2\ 6\ 9 \\ +\ 4\ 3\ 4 \\ \hline \end{array} \rightarrow \begin{array}{r} ^{1}\quad \\ 2\ 6\ 9 \\ +\ 4\ 3\ 4 \\ \hline \mathbf{3} \end{array} \rightarrow \begin{array}{r} ^{1\ 1}\quad \\ 2\ 6\ 9 \\ +\ 4\ 3\ 4 \\ \hline \mathbf{7}\ \mathbf{0}\ 3 \end{array}$

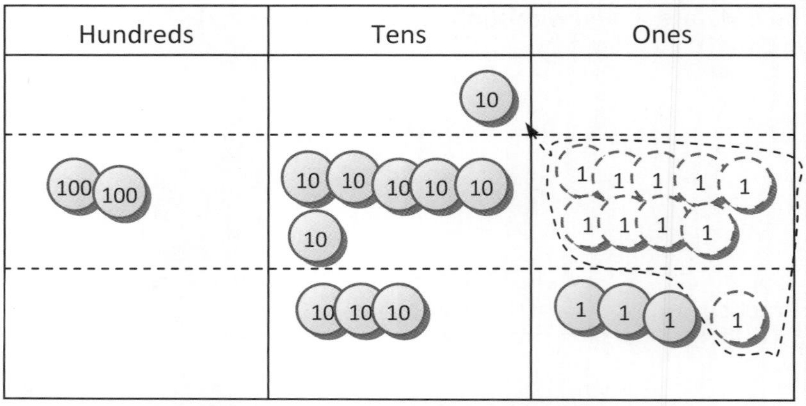

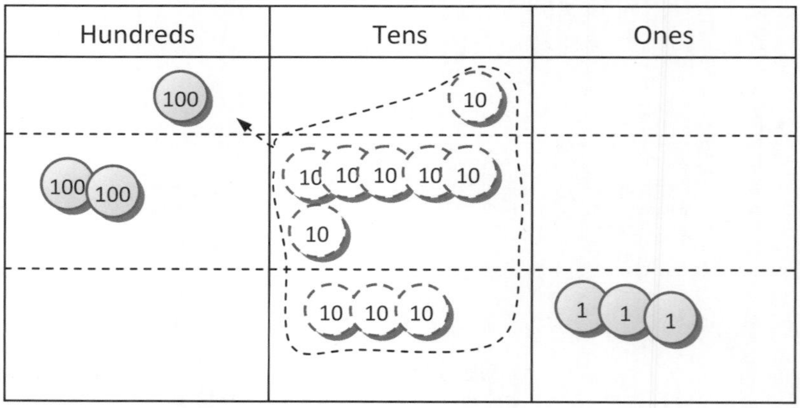

Discussion	Text pp. 39-40
Write the problems on the board. Have students use place-value discs. Guide them in the steps.	11. 248 + 75 = **323** 13. 237 + 184 = **421**
Assessment	**Text p. 40**
	12. (a) 265 + 69 = **334** (b) 493 + 28 = **521** (c) 684 + 19 = **703** 14. (a) 178 + 443 = **621** (b) 204 + 398 = **602** (c) 465 + 135 = **600**
Practice	WB Exercise 18, pp. 46-47, problems 1-2

Enrichment

Have students find the answer to some problems with missing addends, such as the following examples.

```
  2 8 4       2 4 9       2 □ 3       6 □ □
+ □ 7 □     + 1 □ 0     + 2 2 □     + 1 3 6
  9 5 4       4 3 9       4 3 7       □ 6 2
```

Group Game

Material: Number cards 1-9, 4 sets per group.

Procedure: Deal 6 cards to each player. Each player must arrange the cards into two 3-digit numbers and add these together. The player with the lowest total wins.

After a few hands, you can stop the game and discuss ways to arrange the digits into the two numbers. In order to get the lowest total, the two lowest numbers need to be used for the hundreds, the next two lowest numbers for the tens, and the greatest two numbers for the ones. You can have students experiment to see if it matters which number gets which of the two digits.

For example, the cards drawn are 5, 1, 3, 5, 4, and 9.

1 and 3 are lowest so they should be used for hundreds. 4 and 5 should be used for tens, and 5 and 9 for ones.

```
  1 1
  1 4 5
+ 3 5 9
  5 0 4
```

Exercise 18	
1.	2. 301 540 764
C 820 D 325 E 901 G 501	642 700 816
H 373 L 640 T 902 Z 860	830 723 915
	615 702 927
46	47

2.4f Word Problems

Objectives

- Add three numbers.
- Solve word problems.

Note

If you want to give your students more practice with adding 3 numbers, take an extra day with this lesson.

Add columns of numbers	
Write the expression **362 + 194 + 276**. You can show the three numbers with place-value discs or have students use discs and then guide them in solving the problem. Show the results using the written problem. Add the ones first, and rename as tens and ones if there are more than 9 ones. 2 ones + 4 ones + 6 ones = 12 ones = 1 ten 2 ones. We still have tens to add, so write the 1 ten above the tens. The ones is the total ones, so we write that below the line. Then add the tens. 1 ten + 6 tens + 9 tens + 7 tens = 23 tens = 2 hundreds 3 tens. Write the 2 hundreds above the hundreds. Then add the hundreds and write the answer below the line. Ask students to look at the written problem again. Ask them if they see an easier way to add ones without the discs. We can add in any order. Since 4 and 6 make a 10, add that first. Then it is easy to add 2 to 10. Ask them if they see any way to make a hundred with the tens column. They can add 1 ten to 9 tens first.	**1** 3 6 2 1 9 4 + 2 7 6 **2** **2** 1 3 6 2 1 9 4 + 2 7 6 **3** 2 2 1 3 6 2 1 9 4 + 2 7 6 **8** 3 2
Keep the previous problem on the board. Write the expression 166 + 374 + 292 and have students find the answer. See if they realize the answer is the same as in the previous problem. Point out that the 3-digit numbers themselves are different, but the ones, tens, and hundreds are the same.	2 1 1 6 6 3 7 4 + 2 9 2 8 3 2
Add	**Text p. 40**
Task 15: Write the problem on the board. Students should be able to solve it without place-value discs. Use place-value discs if needed. Guide them with the steps. In this example, adding the ones gives 2 tens and 3 ones. Adding the tens, including the 2 tens, gives 1 hundred 7 tens. Task 16: See if students can solve these on their own.	15. 186 + 249 + 38 = **473** 16. (a) 172 + 487 + 74 = **733** (b) 209 + 145 + 567 = **921**

Discussion

Problem 10: Write the problem on the board, or a similar one using objects or situations the students are familiar with. Have them determine which information they need to solve part (a). They can use the answer to part (a) to solve part (b). You can use number bonds.

Point out that there are three parts to this problem, the number sold in the morning, the number sold in the afternoon, and the number left. We could also solve part (b) by adding the three parts together.

Text p. 41, Practice 2C

10. (a) 46 + 28 = **74**
    He sold 74 cream puffs.
    (b) 74 + 16 = **90**
    (or 46 + 28 + 16 = 90)
    He had 90 cream puffs at first.

Sold in morning 46
Sold in afternoon 28
Amount left 16
Total sold 74
Amount he had 90

Assessment

Write problem 3 from Practice 2C and problems 3-5 from Practice 2D on the board horizontally and have students copy them vertically and solve them.

Write word problems 6 and 8 from Practice 2D, or similar problems, on the board and have students solve them individually and then share their solutions.

Text pp. 41-42, Practices 2C and 2D

3. (a) 44 + 56 = **100**
   (b) 58 + 42 = **100**
   (c) 74 + 26 = **100**
3. (a) 745 + 108 = **853**
   (b) 829 + 122 = **951**
   (c) 667 + 227 = **894**
4. (a) 490 + 139 = **629**
   (b) 584 + 250 = **834**
   (c) 876 + 19 = **895**
5. (a) 293 + 60 + 24 = **377**
   (b) 339 + 104 + 40 = **483**
   (c) 224 + 106 + 320 = **650**
6. 169 + 71 = **240**
   He has 240 stamps now.
8. 102 + 86 + 40 = **228**
   228 people were at the concert.

Practice

WB Exercise 18, p. 48, problems 3-5
WB Exercise 19, pp. 49-50

Exercise 18

3. $82 + $139 = $221
   $221

4. $393 + $438 = $831
   $831

5. 468 + 156 = 624
   624

48

Exercise 19

1. 327 735

691 731 308

815 725

GOOD
MORNING

49

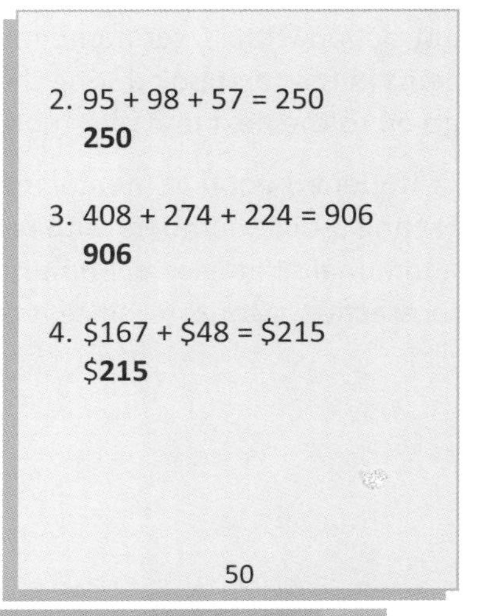
2. 95 + 98 + 57 = 250
   250

3. 408 + 274 + 224 = 906
   906

4. $167 + $48 = $215
   $215

50

2.5 Subtraction With Renaming

Objectives

- Subtract ones or tens from a 3-digit number using mental math strategies.
- Subtract within 1000.
- Solve word problems.

Material

- Place-value discs and charts
- Base-10 blocks for board
- Number cube 4-9, 1 per group
- Number cards 0-9, 4 sets per group
- Mental Math 16-18 (appendix)
- Appendix p. a20

Prerequisites

Students should have a good knowledge of place value and of the basic addition facts within 20. For the lesson on mental math, they should be familiar with the "subtract from a ten" strategy as reviewed in lesson 1.1d.

Notes

In *Primary Mathematics* 1B students learned to subtract a 1-digit number from a 2-digit number mentally when renaming occurs. This will be reviewed here and extended to subtracting ones or tens from a 3-digit number. Mental math strategies are often based on the base-10 aspect of our number system, and are important for students to learn in order to gain good number sense. Since they vary depending on the problem, or different strategies can be applied to the same problem, they can be challenging for some students. If mental math is a struggle for your students, or they did not use earlier levels of *Primary Mathematics*, you can save the first lesson in this section for later and allow students to use the subtraction algorithm, or just do the part of the lesson involving tens minus ones and 2-digit numbers minus 1-digit numbers. Then introduce the concepts in this lesson over a longer period of time as you continue with the curriculum.

In this part students will subtract a number from another number within 1000 where renaming occurs using the standards algorithm. They will first subtract when tens need to be renamed as ones, then when hundreds need to be renamed as tens, and finally when both tens and hundreds need to be renamed.

Show students the algorithm, step by step, with a place-value chart and discs, and relate these steps to the written problem. These steps are shown on the next page for a subtraction problem that involves renaming of both tens and hundreds. In this example, a little 1 is written next to the 2 to indicate 12 ones. In the example in the textbook on p. 43, the ones are crossed out and a 12 written above them. Either method is acceptable, with the method in the textbook being a bit more explicit.

When discussing the steps, always use the place-value name with the digit. In the example in the next page, we are subtracting 8 *tens* from 12 *tens*, not 8 from 12.

Subtraction with larger numbers can be time-consuming for students at first. If there is not time to do all the problems in the textbook during the lesson, save some of them and give students several problems each day as you go on to the next units for more continuous practice.

If a student is good at mental math, he or she may be able to come up with mental techniques for subtracting 3-digit numbers with renaming. Some specific strategies will be taught in *Primary Mathematics* 2B, and additional strategies in *Primary Mathematics* 3. However, all students should be thoroughly familiar with the subtraction algorithm, since it can be used in all cases.

Hundreds	Tens	Ones
100 100 100 100 100 100 100 100	10 10 10	1 1

$$\begin{array}{r} 8\ 3\ 2 \\ -\ 4\ 8\ 5 \\ \hline \end{array}$$

Cannot subtract 5 ones from 2 ones. Rename a ten as 10 ones.
3 tens 2 ones = 2 tens 12 ones

Hundreds	Tens	Ones
100 100 100 100 100 100 100 100	10 10 10	1 1 1 1 1 1 1 1 1 1 1 1

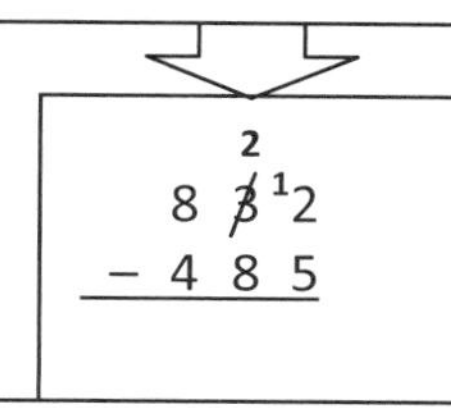

Subtract the ones.
12 ones – 5 ones = 7 ones
Write the answer in the ones place.

Hundreds	Tens	Ones
100 100 100 100 100 100 100 100	10 10	1 1 1 1 1 1 1

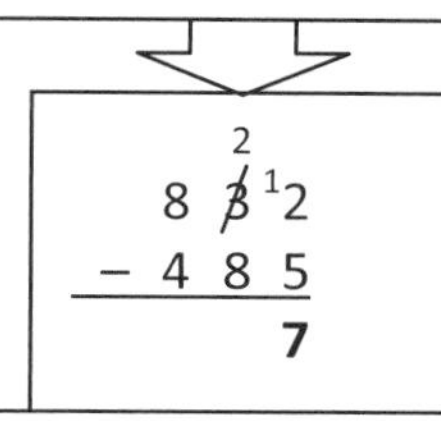

Cannot subtract 8 tens from 2 tens. Rename a hundred as 10 tens.
8 hundreds 2 tens = 7 hundreds 12 tens

Hundreds	Tens	Ones
100 100 100 100 100 100 100 100	10 10 10 10 10 10 10 10 10 10 10 10	1 1 1 1 1 1 1

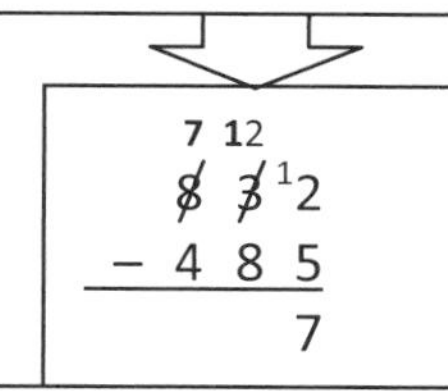

Subtract the tens.
12 tens – 8 tens = 4 tens
Write the answer in the tens place.

Hundreds	Tens	Ones
100 100 100 100 100 100 100	10 10 10 10	1 1 1 1 1 1 1

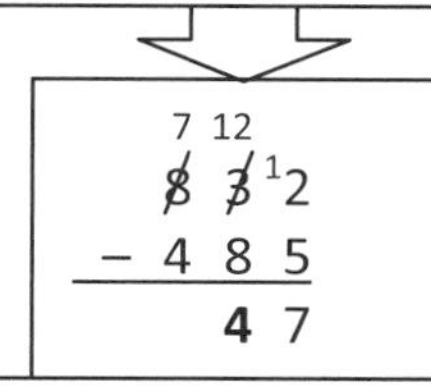

Subtract the hundreds.
Write the answer in the hundreds place.

Hundreds	Tens	Ones
100 100 100	10 10 10 10	1 1 1 1 1 1 1

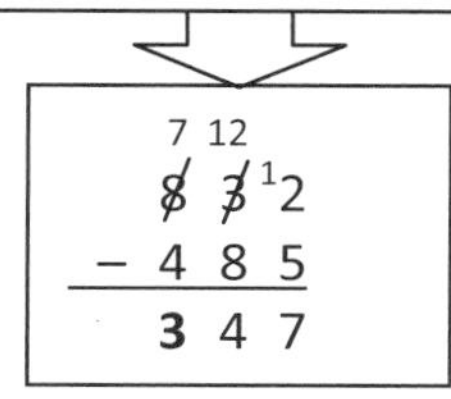

2.5a Subtract Ones or Tens from a 3-Digit Number

Objectives

Subtract ones or tens from a 3-digit number using mental math strategies.

Note

In this lesson students will subtract ones or tens from a 3-digit number. The mental math strategies are an extension of subtracting ones mentally, even when there is renaming, as in 14 – 9, which students have already learned. Students can use either a "subtract from a ten" strategy, or use subtraction facts. Those students who have difficulty memorizing subtraction facts, or who forget a fact, will be able to use the "subtract from a ten" strategy in order to make accurate calculations.

Mental math can be challenging for some students. If students struggle with mental math, they can always fall back on the standard subtraction algorithm, which they will learn in the next lesson. Students can also solve the problems by counting down in the correct place and keeping track with fingers. For example, 432 – 70: Count down by tens from 432: 42 (tens), 41, 40, 39, 38, 37, 36 → 362.

You can draw number bonds to illustrate the thinking process. Students are expected to eventually do these problems without needing to draw number bonds.

Illustrate any of the problems in this lesson with place-value discs, as needed.

Subtract ones	
Write the expression **15 – 7** on the board. Discuss different methods for finding the answer: ⇒ Subtract from ten. ⇒ Subtract 5 from 15, then 2 more. ⇒ "Count up" from 7 first to 10, then to 15. ⇒ Remember the math fact 15 – 7 = 8.	15 – 7 15 – 7 = 5 + 3 = 8 /\\ 5 10 – 7
Have students solve **10 – 7** and then write **40 – 7**. You can illustrate this problem by setting out four ten-discs or four dimes. Tell students you want to buy something that costs 7 cents. What would you use to pay for it? (A dime.) What change would you get? (3 cents.) How much money do you now have? (3 dimes and 3 pennies, 33 cents.) There is one less dime, or ten.	10 – 7 = 3 40 – 7 40 – 7 = 30 + 3 = 33 /\\ 30 10
Write the expression **45 – 7**. Ask students if there are enough ones to take 7 from. There are not, so we need to involve the tens. Discuss some strategies for finding the answer. ⇒ Subtract 7 from 40, as before, and add back in the 5 ones. You can illustrate this strategy with making change, using number discs or coins. If we have 4 dimes and 5 pennies, then to buy something that costs 7 cents we would have to give a dime and get 3 cents back. So we now have 3 dimes, 3 pennies, and the 5 pennies we started with. ⇒ Subtract 5, and then 2 more (the difference between 7 and 5). ⇒ Remember that 15 – 7 = 8.	45 – 7 45 – 7 = 33 + 5 = 38 /\\ 5 40 45 – 7 = 45 – 5 – 2 = 38 /\\ 5 2 (7 – 5 = 2, 40 – 2 = 38) 45 – 7 = 30 + 8 = 38 /\\ 30 15

Write the expression **645 – 7**. Since subtracting 7 will not change the hundreds, we can subtract as we would 45 – 7, but the answer must include the hundreds.	645 – 7 645 – 7 = 638 /\\ 600 45
Subtract tens	
Write the expression **400 – 70**. Discuss some strategies for finding the answer. ⇒ Solve this problem in the same way as 40 – 7 by thinking of it as 40 tens – 7 tens. ⇒ Subtract 70 from 100. (10 – 7 = 3 so 100 – 70 = 30) You can illustrate this with making change. If you have 4 dollar bills, how will you pay for something that costs 70 cents?	400 – 70 40 tens – 7 tens = 33 tens 400 – 70 = 300 + 30 = 330 /\\ 300 100
Write the expression **450 – 70**. Lead students to see that 45 tens – 7 tens can be solved by finding the answer to 45 – 7.	450 – 70 = 380
Extra: Write the expression **451 – 70**. We can simply find 450 – 70, then add the ones.	451 – 70 = 381
Assessment	**Text p. 44**
Task 1: Have students do these, in the order (a), (d), (b), (e), (c), (f). You can write them in that order on the board.	1. (a) 10 – 6 = **4** (d) 100 – 60 = **40** (b) 11 – 6 = **5** (e) 110 – 60 = **50** (c) 41 – 6 = **35** (f) 410 – 60 = **350**
Practice	Mental Math 16-17

2.5b Rename Tens

Objectives

- Subtract within 1000 by renaming tens.

Note

Because the problems in the workbook are mixed, the pages will be assigned out of order so that all the problems on the assigned workbook page for this lesson involve only renaming in the tens, for the next lesson only renaming in the hundreds, and so on. If you prefer to assign workbook pages in the same order they are encountered in the workbook, you will have to wait until after lesson 2.5e, since some problems in Exercise 20 involve renaming in both tens and hundreds, or change the numbers in those problems.

Rename tens	
Write the problems shown here on the board and ask students to fill in the blanks. Illustrate renaming with place-value discs, if needed. Tell students we are ungrouping a ten into ones, or trading in a ten for ones, or renaming a ten as ones. Ask if we have the same number of objects if we have them in 6 groups of ten and two groups of one, or if we have them in 5 groups of ten and 12 groups of one. We do. If your students are confused, use base-10 blocks instead, since it is easier to visualize the equivalency.	62 = 6 tens ______ ones 62 = 5 tens ______ ones 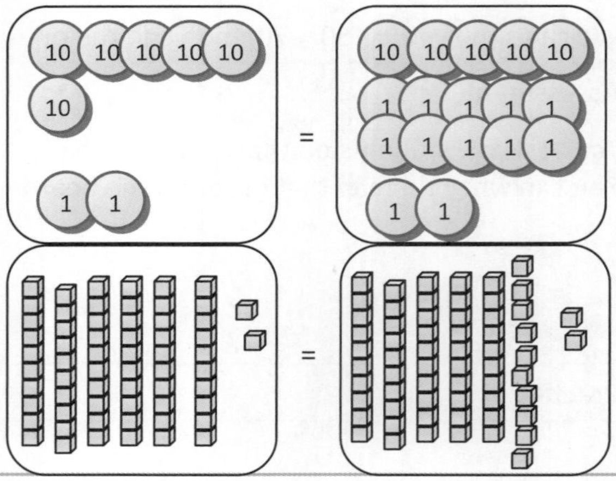 10 10 10 10 10 10 1 1 = 10 10 10 10 10 1 1 1 1 1 1 1 1 1 1 1 1
Write **62** on the board and show the number with place-value discs. Then write **62 – 3** and ask students how we can take away 3 from ones rather than from a ten. To do so, we need to rename one of the tens first. Write the problem vertically. Tell students we can show what just did with the place-value discs by crossing out the 62 and writing 5 tens and 12 ones above it. Or we can just cross out the 6, write 5 ones above it, and put a little 1 next to the 2 to show 12 ones. Then we show that we took away 3 ones by writing the answer to 12 – 3 under the line in the ones place. We also write 5 under the line to show there are 5 tens in the final answer to 62 – 3.	62 – 3 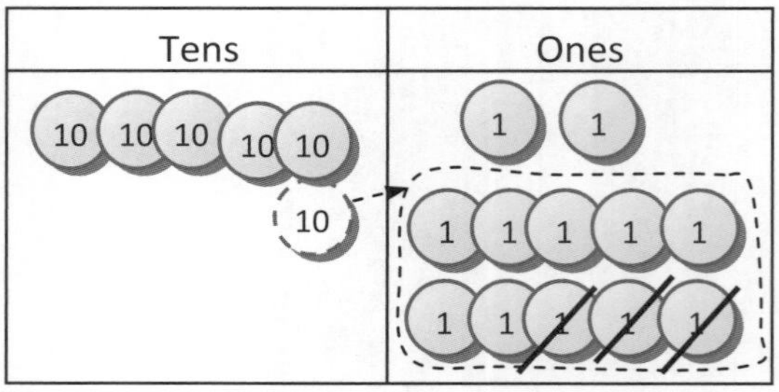 Tens: 10 10 10 10 10 (10) / Ones: 1 1, 1 1 1 1 1 1 1 1 1 1 5 12 over ~~6~~ ~~2~~, – 3, 5 9 or 5 over ~~6~~ [1]2, – 3, 5 9
Show 62 with discs again. Write **62 – 43** vertically. Ask students to subtract the ones and then the tens. Again, show how this is recorded on the written problem. We subtract 4 from the 5 that we have written above in the tens place, since that is how many tens we have left, and write the answer under the line. (This is the problem on p. 43 of the textbook.)	62 – 43 5 over ~~6~~ [1]2, – 4 3 → 5 over ~~6~~ [1]2, – 4 3, **9** → 5 over ~~6~~ [1]2, – 4 3, **1** 9

Have students show 762 with their place-value discs. Write the expression **762 – 43** vertically. Show how we record each step on the written problem. Repeat with **762 – 543**.	$\begin{array}{r} 7\ \overset{5}{\cancel{6}}\ {}^{1}2 \\ -\quad 4\ 3 \\ \hline 7\ 1\ 9 \end{array}$ $\qquad$ $\begin{array}{r} 7\ \overset{5}{\cancel{6}}\ {}^{1}2 \\ -\ 5\ 4\ 3 \\ \hline 2\ 1\ 9 \end{array}$
Subtract	**Text pp. 44-45**
Write the problems on the board. Have students use place-value discs. Guide them in recording the steps.	3. 243 – 18 = **225** 5. 452 – 134 = **318**
Assessment	**Text pp. 44-45**
	2. (a) 30 – 6 = **24** (b) 41 – 9 = **32** (c) 52 – 13 = **39** (d) 63 – 35 = **28** (e) 74 – 48 = **26** (f) 86 – 58 = **28** 4. (a) 354 – 9 = **345** (b) 480 – 7 = **473** (c) 562 – 34 = **528** (d) 690 – 45 = **645** (e) 720 – 18 = **702** (f) 833 – 29 = **804** 6. (a) 441 – 227 = **214** (b) 553 – 228 = **325** (c) 764 – 506 = **258** (d) 470 – 256 = **214** (e) 625 – 118 = **507** (f) 830 – 724 = **106**
Reinforcement	
Show students how to check the answer to a subtraction problem with addition. Since we subtract a part from the whole to get a part, if we add the two parts together, we should get the whole again. Have students check the answers of some of the subtraction problems they have already solved by adding the answer to the number being subtracted. They can either rewrite the problem, or add up from the bottom.	$\begin{array}{r} 5\ \overset{1}{\cancel{2}}\ {}^{1}2 \\ -\ 2\ 1\ 6 \\ \hline 3\ 0\ 6 \end{array}$ Check: $\begin{array}{r} \underline{5\ 2\ 2} \\ +\ 2\ 1\ 6 \\ 3\ 0\ 6 \end{array} \Uparrow$
Practice	WB Exercise 20, p. 51, problem 1

Exercise 20

1. C 15 D 38 E 37

I 39 M 4 N 6

O 46 S 28 T 26

IT DOES NOT
COME TO ME

51

2.5c Rename Hundreds

Objectives

- Subtract within 1000 by renaming hundreds.

Note

Students should realize that the process for each place repeats itself; the difference is only in the place value of the digit.

More capable students will not need to spend as much time looking at each place separately or need as much practice so you can just start with the example in the textbook and combine this lesson with 2.5e, and then combine lessons 2.5d and 2.5f.

Rename hundreds	
Write **428 – 53** on the board and show 428 with place-value discs. Then rewrite 428 – 53 vertically. Ask students how we can subtract ones (just remove 3 ones). Then ask how we can subtract tens. To remove 4 tens, they need to replace a hundred with 10 tens. Show students how we record these steps on the written problem: First we subtract 3 from 8 and write the number of ones below the line. Then, we need to subtract 5 tens. There are not enough tens, so we cross out the 4 hundreds and rename it as 3 hundreds 10 tens. We now have 12 tens. We show that by writing a 1 next to the 2 tens (or crossing it out and writing 12 above it in the tens place). Then we find the answer to 12 tens – 5 tens, which is 7 tens, and write that below the line, since it is the final number of tens. Then we can write the final number of hundreds, which is 3 hundreds, below the line.	428 – 53 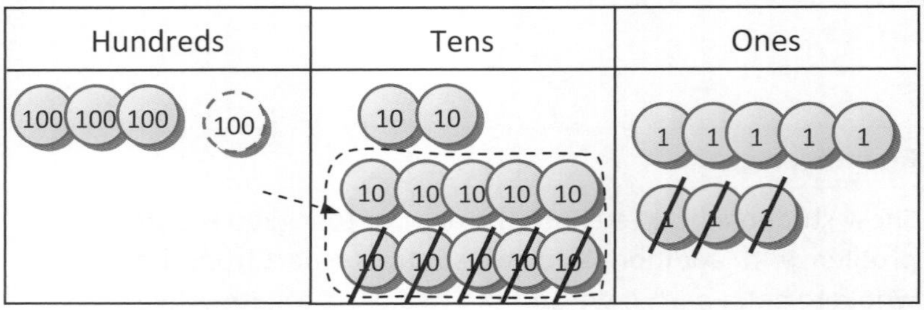 Hundreds: 100 100 100 (100) · Tens: 10 10, 10 10 10 10 10, 10 10 10 10 10 · Ones: 1 1 1 1 1, 1 1 1 $\begin{array}{r} 4\;2\;8 \\ -\;\;\;5\;3 \\ \hline 5 \end{array} \rightarrow \begin{array}{r} ^{3}\;\;\;\;\;\; \\ \not{4}\;{}^{1}2\;8 \\ -\;\;\;5\;3 \\ \hline 5 \end{array} \rightarrow \begin{array}{r} ^{3}\;\;\;\;\;\; \\ \not{4}\;{}^{1}2\;8 \\ -\;\;\;5\;3 \\ \hline 3\;7\;5 \end{array}$
Write the expression **428 – 153** and have students tell you the steps. The steps are the same, except that this time we need to subtract hundreds.	428 – 153 $\begin{array}{r} 4\;2\;8 \\ -\;1\;5\;3 \\ \hline 5 \end{array} \rightarrow \begin{array}{r} ^{3}\;\;\;\;\;\; \\ \not{4}\;{}^{1}2\;8 \\ -\;1\;5\;3 \\ \hline 5 \end{array} \rightarrow \begin{array}{r} ^{3}\;\;\;\;\;\; \\ \not{4}\;{}^{1}2\;8 \\ -\;1\;5\;3 \\ \hline 2\;7\;5 \end{array}$
Repeat with **526 – 464**. This time there will be no hundreds left.	526 – 464 $\begin{array}{r} 5\;2\;6 \\ -\;4\;6\;4 \\ \hline 2 \end{array} \rightarrow \begin{array}{r} ^{4}\;\;\;\;\;\; \\ \not{5}\;{}^{1}2\;6 \\ -\;4\;6\;4 \\ \hline 2 \end{array} \rightarrow \begin{array}{r} ^{4}\;\;\;\;\;\; \\ \not{5}\;{}^{1}2\;6 \\ -\;4\;6\;4 \\ \hline 6\;2 \end{array}$

Subtract	Text pp. 45-46
Write the problems on the board. Have students copy them in a vertical format. Let students use place-value discs. Guide them	7. 729 – 64 = **665** 9. 538 – 293 = **245**
Assessment	**Text pp. 45-46**
	8. (a) 348 – 76 = **272** (b) 409 – 38 = **371** (c) 516 – 54 = **462** (d) 707 – 61 = **646** (e) 620 – 80 = **540** (f) 139 – 83 = **56** 10. (a) 617 – 247 = **370** (b) 308 – 140 = **168** (c) 705 – 492 = **213** (d) 807 – 486 = **321** (e) 634 – 284 = **350** (f) 920 – 840 = **80**
Practice	WB Exercise 21, p. 54, problem 1

Group Game: Roll 0

Material: Number cube 4-9, 1 per group. Paper with three columns for ones, tens, and hundreds.

Procedure: Each player writes 500, rolls the number cube once, and writes the number down in either the ones or the tens column. If it is in the tens column, the player writes a 0 in the ones column and subtracts the tens. Each rolls the cube a second time, decides whether it is to be tens or ones, and subtracts it from the previous number. The play continues until each player has rolled the cube 10 times. The goal is for the final difference to be as close to 0 as possible.

	5 0 0
1st roll, 6	– 6 0
	4 4 0
2nd roll, 5	5 0
	3 9 0
3rd roll, 9	9 0
	3 0 0
4th roll, 8	8 0
	2 2 0
5th roll, 7	7 0
	1 5 0
6th roll, 8	8 0
	7 0
7th roll, 9	9
	6 1
8th roll, 4	4 0
	2 1
9th roll, 4	4
	1 7
10th roll, 6	6
	1 1

Exercise 21

1. A 454 B 154 C 295

D 81 E 522

F 352 G 685 H 774

the treasure is hidden below the rain tree

54

2.5d Word Problems

Objectives

- Subtract within 1000 by renaming tens or hundreds.
- Solve word problems.

Note

You will be discussing two problems where we are comparing two numbers, but in both we are given the difference, rather than having to find the difference. In the first one, we need to find the larger number, so we add. In the second, we need to find the smaller number, so we subtract. The two problems are worded similarly. Just because a problem uses the word "more" the student should not automatically subtract the two numbers. If students have trouble with these problems, have them act them out with number discs, or rephrase the problems with smaller numbers so they can use counters or objects to act out the problems.

Practice subtraction	
Write some or all of the expressions at the right on the board, horizontally. Allow students to attempt to solve them first mentally. Have students rewrite any that they cannot solve mentally, and solve using the subtraction algorithm. Guide them through the steps of some, as needed, and have them work on the rest independently.	56 – 8 (48) 56 – 38 (18) 756 – 38 (718) 756 – 82 (674) 756 – 382 (374) 965 – 423 (542) 429 – 161 (268)
More practice	**Text p. 48, Practice 2E**
You may also use all or some of the problems from Practice 2E, p. 48, particularly if students still need practice in renaming tens. Some students will be able to solve these mentally. Mental math strategies for subtracting 2-digit numbers when renaming occurs will be reviewed in *Primary Mathematics* 2B.	1. (a) 32 (b) 38 (c) 36 2. (a) 27 (b) 29 (c) 25 3. (a) 18 (b) 48 (c) 45 4. (a) 9 (b) 7 (c) 1 5. (a) 5 (b) 5 (c) 5
Word problems	**Text p. 48, Practice 2E**
Problem 6: Write the problem on the board. Ask students what we need to find. (How many storybooks Devi has.) Which child has more storybooks? (Since Devi has more than Ailian, she has more storybooks.) We are comparing two numbers, how many storybooks Ailian has and how many Devi has. Are we finding how many more one is than the other? (No.) We are given how many more Devi has than Ailian. We can use a number bond to show the information in the problem. The two parts are how many books Ailian has and how many more books Devi has. So we need to find how many books Devi has using addition. Problem 7: Write the problem on the board. Ask what we need to find. (How many shells Mary has.) Which child has more shells? (Jenny.) So Mary has fewer shells. Are we told how many fewer? (Yes.) How can we find the number of shells Mary has? (We subtract.)	6. Ailian 18, more 14 → Devi ? 18 + 14 = **32** Devi has 32 books. 7. Jenny 92 → Mary ?, more 9 92 – 9 = **83** Mary collected 83 shells.

Assessment	Text p. 48, Practice 2E
Have students work on the problems individually and then share their solutions. Be sure they notice that in problem 10 the price of the dictionary is written on the picture of the book. You may want to write the problems on the board to be sure students are doing the correct problems. Students will need to read the problems carefully; they are not all subtraction problems just because that is what they are learning in the current unit. Do not require students to draw number bonds if they do not need them to solve the problem. Number bonds are a tool to help students determine which operation they need; students should eventually be able to solve these kinds of simple problems without drawing a visual aid.	8. 84 – 15 = **69** She had 69 T-shirts left. 9. \$92 – \$58 = **\$34** She had \$34 left. 10. \$42 + \$28 = **\$70** She had \$70 at first.
You can have students do some additional problems.	Appendix p. a20
1. Harriet bought 148 bags of birdseed. So far she has used 92 bags. How many bags does she have left?	148 – 92 = 56 She has 56 bags left.
2 Harriet saw a cockatoo for sale at a breeder's. It cost \$750. She had \$680. How much more money does she need to buy the cockatoo?	\$750 – \$680 = \$70 She needs \$70 more.
Practice	WB Exercise 20, pp. 52-53, problems 2-5 WB Exercise 21, p. 55, problems 2-4
Problem 4 on p. 53 in the current printing requires subtracting 132 – 34, which involves renaming in both hundreds and tens. You may want to see how students do on this; it is easy to solve mentally by subtracting 34 and then another 2. Alternately, you can change the first number to 142 or save this page to do after lesson 2.5f.	

Exercise 20

2. A 735 B 343 E 26

L 363 M 333

N 116 R 745 U 540

AN
UMBRELLA

52

3. \$186 – \$38 = \$148
\$148

4. 132 – 34 = 98
98

5. 150 – 43 = 107
107

53

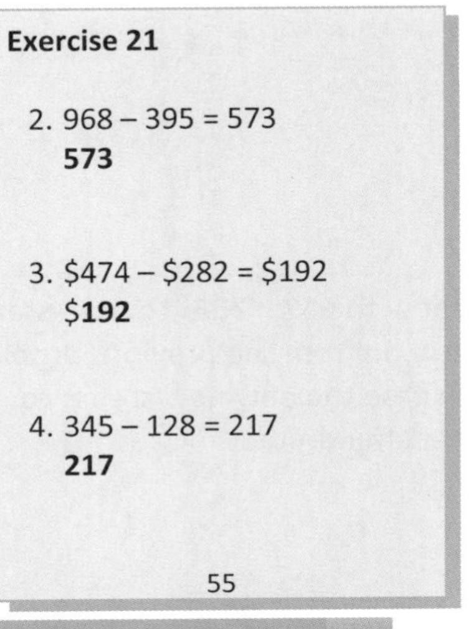

Exercise 21

2. 968 – 395 = 573
573

3. \$474 – \$282 = \$192
\$192

4. 345 – 128 = 217
217

55

Rename Tens and Hundreds

Objectives

- Subtract within 1000 by renaming both tens and hundreds.

Note

This lesson covers subtraction with renaming in both tens and hundreds. Some students may not need to spend as much time practicing renaming in specific places and this lesson might be too redundant. You can combine with the next lesson or include part of the next lesson on subtracting from hundreds.

Rename tens and hundreds	
Write the expression **542 – 6** on the board and show the number 542 with place-value discs. Ask students how we can take away 6 ones. Show how this is recorded, as in previous lessons.	542 – 6
Repeat with **542 – 56**. If students are following along with their own discs, you can have them carry on from the answer in the last problem, or start over again at the beginning, by showing 542 again, and subtracting ones again. There are not enough tens, so a hundred has to be renamed as 10 tens before we can take 5 tens away. Again, show how the steps are recorded on the written problem.	542 – 56 Hundreds / Tens / Ones
Repeat with **542 – 256**. You can start over, or carry on from the previous problem, in which case the only new step is to subtract hundreds.	542 – 256

Repeat with a new problem, **432 – 138**. Initially there are enough tens to subtract 3 tens, but after we subtract ones, there are not enough tens, and we need to then rename a hundred. You can have students try this problem by starting with the hundreds to show why sometimes it is easier to use the algorithm instead of mental math. The hundreds and tens will have to changed when we get to ones. So the answer for each place value is not final. By starting with ones, we don't have to change the answer.	432 – 138 3 12 ~~4~~ ~~3~~ 12 – 1 3 8 2 9 4 4 3 2 – 1 3 8 3 0 ? → 302 – 8
Subtract	**Text pp. 46-47**
Let students use place-value discs for these, if needed.	11. 421 – 68 = **353** 13. 453 – 267 = **186**
Assessment	**Text p. 47**
	12. (a) 322 – 47 = **275** (b) 430 – 55 = **375** (c) 631 – 78 = **553** 14. (a) 512 – 149 = **363** (b) 640 – 276 = **364** (c) 623 – 246 = **377**
Practice	WB Exercise 22, pp. 56-57, problems 1-2

Group Game

Material: Number cards 1-9, 4 sets per group.

Procedure: Deal 6 cards to each player. Each player must arrange the cards into two 3-digit numbers and subtract the smaller number from the greater number. The player with the lowest total wins.

Exercise 22

1. B 253 S 568 D 759

E 75 A 217 O 649

O 579 K 489 R 277

READ BOOKS

56

2. 41 198 269

149 195

512 298 77

78 374

57

2.5f More Subtraction

Objectives

- Subtract within 1000 when there are no tens.
- Solve word problems.

Note

This lesson fist covers subtraction from hundreds or where renaming occurs over two places and then continues with a practice. If students need more practice with subtraction, provide a variety of subtraction problems rather than proceeding directly to word problems.

Rename hundreds as 9 tens 10 ones	
Display a hundred-disc and ask students how many tens are in a hundred. Replace the disc with 10 tens and then ask students how many ones are in a ten. Replace 1 ten with 10 ones. Ask students if we still have 100. We do. So 1 hundred = 9 tens and 10 ones. If students have trouble with this, use base-10 blocks. It is easier to see from a hundred-flat that it is the same as 9 tens and 10 ones.	100 = 10 10 10 10 10 10 10 10 10 + 1 1 1 1 1 1 1 1 1 1 100 = 9 tens + 10 ones
Show 500 with place-value discs. Then write the expression **500 – 4**. Ask students how we can subtract 4. We can't rename tens; there are not any tens. So we need to rename hundreds. We can rename 1 hundred as 10 tens, but we still don't have ones. So we need to rename one of the tens as 10 ones before we can subtract. Show the steps on the written problem. Tell students that we can also simply rename a hundred as 9 tens and 10 ones in one step.	500 – 4 $\begin{array}{r} 5\;0\;0 \\ -\;\;\;\;\;4 \\ \hline \end{array} \rightarrow \begin{array}{r} \overset{4}{\cancel{5}}\;{}^{1}0\;0 \\ -\;\;\;\;\;\;4 \\ \hline \end{array} \rightarrow \begin{array}{r} \overset{4}{\cancel{5}}\;\overset{9}{{}^{1}\cancel{0}}\;{}^{1}0 \\ -\;\;\;\;\;\;4 \\ \hline \mathbf{4\;9\;6} \end{array}$ $\begin{array}{r} 5\;0\;0 \\ -\;\;\;\;\;4 \\ \hline \end{array} \rightarrow \begin{array}{r} \overset{4}{\cancel{5}}\;\overset{9}{\cancel{0}}\;{}^{1}0 \\ -\;\;\;\;\;\;4 \\ \hline \mathbf{4\;9\;6} \end{array}$
Write the expression **603 – 137**, show 603 with place-value discs, and have students first subtract the ones, then the tens, and then the hundreds. To subtract the ones, they need to rename a hundred. Show them how we record the steps on the written problem.	603 – 137 $\begin{array}{r} 6\;0\;3 \\ -\;1\;3\;7 \\ \hline \end{array} \rightarrow \begin{array}{r} \overset{5}{\cancel{6}}\;\overset{9}{\cancel{0}}\;{}^{1}3 \\ -\;1\;3\;7 \\ \hline \end{array} \rightarrow \begin{array}{r} \overset{5}{\cancel{6}}\;\overset{9}{\cancel{0}}\;{}^{1}3 \\ -\;1\;3\;7 \\ \hline \mathbf{4\;6\;6} \end{array}$
Subtract	**Text p. 47**
Write the problem on the board. Students can use place-value discs. Guide them in recording their answers.	15. 300 – 28 = **272**

Assessment	Text p. 47
	16. (a) 400 – 38 = **362** (b) 700 – 276 = **424** (c) 402 – 337 = **65**
Practice	**Text p. 49, Practice 2F**
Students may be able to do these independently. They can use number bonds as needed, but do not require them for every problem. A suggested number bond is given for problem 8. Problem 10: This is essentially a 2-step problem, with the first step provided. In *Primary Mathematics* 3A, students will get problems like this where only (b) is asked.	1. (a) 320 (b) 432 (c) 540 2. (a) 77 (b) 308 (c) 425 3. (a) 162 (b) 207 (c) 207 4. (a) 391 (b) 394 (c) 394 5. (a) 177 (b) 416 (c) 77 6. 320 – 180 = **140** 140 chairs are not new. 7. 224 + 298 = **522** There are 522 buttons althogether. 8. 105 – 87 = **18** He gave away 18 cards. 9. 620 – 465 = **155** She needs 155 more beads. 10. (a) 304 - 46 = **258** There are 258 boys. (b) 304 + 258 = **562** There are 562 children in the school.
Practice	WB Exercise 22, p. 58, problems 3-5 WB Exercise 23, pp. 59-60

Exercise 22

3. 415 – 158 = 257
257

4. 250 – 174 = 76
76

5. $620 – $565 = $55
$55

58

Exercise 23

1.

268 I	138 N		
362 T	546 H	26 E	
659 W	26 E	485 L	485 L

59

2. 700 – 369 = 331
331

3. 504 – 286 = 218
218

4. 207 – 179 = 28
28

60

2.5g Practice and Review

Objectives

- Practice addition and subtraction within 1000.
- Review all topics.

Note

You may want to spend more than one day on this practice and review, or you may want to save some of the problems in the textbook practices for another day to provide more continuous review during the next units and assign a few problems whenever time permits.

Practice	**Text pp. 50-51, Practices 2G and 2H**	
Students can do these problems individually and share their answers, or you can do some of the word problems as a class.	Practice 2G 1. (a) 79 (b) 79 (c) 99 2. (a) 59 (b) 40 (c) 6 3. (a) 108 (b) 80 (c) 109 4. (a) 33 (b) 35 (c) 5 5. (a) 100 (b) 101 (c) 107 6. 98 – 39 = **59** Kelly has 59 postcards. 7. 82 + 24 = **106** He has 106 ducks. 8. (a) Team B (b) 95 – 79 = **16** Team B scored 16 more points. 9. (a) 26 + 9 = **35** Paul is 35 years old. (b) 35 + 8 = **43** Mary is 43 years old.	Practice 2H 1. (a) 210 (b) 413 (c) 320 2. (a) 199 (b) 288 (c) 699 3. (a) 301 (b) 260 (c) 47 4. (a) 284 (b) 555 (c) 607 5. (a) 358 (b) 337 (c) 96 6. 427 + 278 = **705** There are 705 cards in the two lots. 7. 152 – 35 = **117** There are 117 chairs in the hall. 8. $220 – $186 = **$34** The watch cost $34. 9. 140 – 23 = **117** 117 children passed the test. 10. (a) $212 – $144 = **$68** The calculator costs $68. (b) The watch costs more. $144 – $68 = **$76** It costs $76 more.
Enrichment	Mental Math 18	
Write the problems on the board. See if students can find the answers without evaluating both sides.	Write >, <, or = in each circle. 470 + 50 ◯ 500 (>) 342 + 50 ◯ 350 + 42 (=)	 450 + 300 ◯ 450 + 30 (>) 459 + 50 ◯ 459 + 70 (<)
Discuss the problems at the right, or similar problems. See if students can recognize from the wording whether to add or subtract.	457 is 42 more than ____ ____ is 35 more than 901. 139 more than _____ is 682	(457 – 42 = 415) (901 + 35 = 936) (682 – 139 = 543)
Practice	WB Exercise 24, pp. 61-62 WB Review 1, pp. 63-66	

Enrichment

Have students find the answer to some problems with missing digits, such as the ones below.

$$\begin{array}{r} 3\ 0\ 2 \\ -\ 1\ \square\ 8 \\ \hline \square\ 3\ 4 \end{array} \qquad \begin{array}{r} 4\ \square\ 2 \\ -\ 3\ 1\ \square \\ \hline 1\ 3\ 4 \end{array} \qquad \begin{array}{r} 4\ \square\ \square \\ -\ 3\ 7\ 8 \\ \hline 5\ 4 \end{array}$$

Group Game

Material: Number cards 0-9, 4 sets per group.

Procedure: Each group selects a dealer. The dealer shuffles the cards and turns over the first three cards. The first card is the hundreds, the second card the tens and the third card the ones of what becomes the target number. The dealer then deals six cards to each player. The players arrange their cards into two 3-digit numbers so that the difference will be as close to the target number as possible. The winner is the player whose difference is the closest.

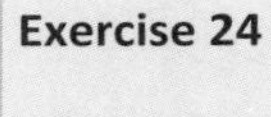

Exercise 24

1. A 81 C 775 H 1000

I 327 M 378 R 530

S 277 S 723 T 638

MERRY CHRISTMAS

61

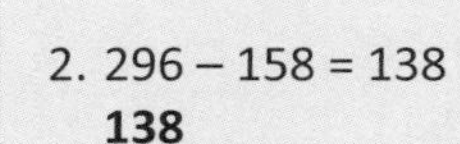

2. 296 – 158 = 138
138

3. 150 – 78 = 72
72

4. 930 – 845 = 85
85

62

Review 1

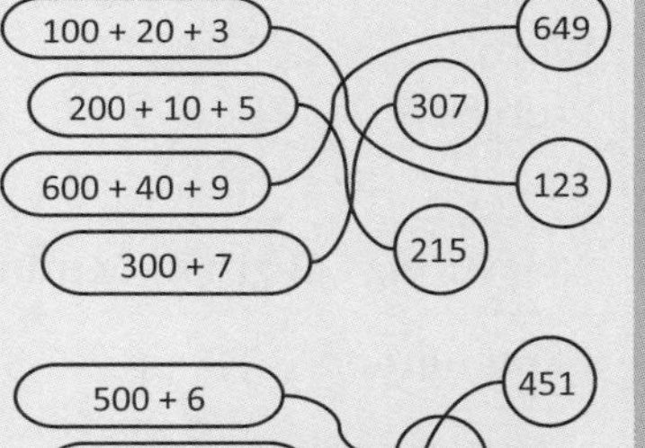

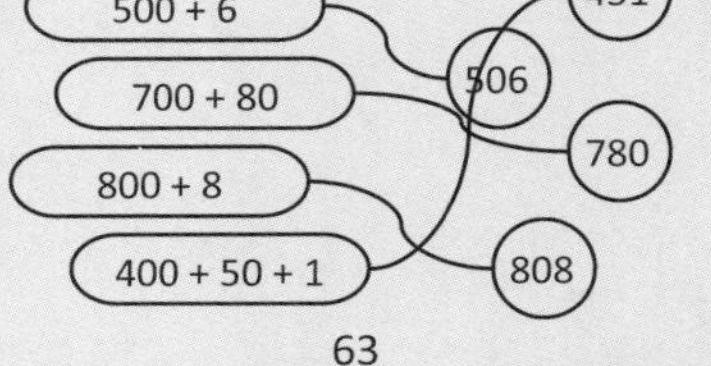

63

2. (a) 89
(b) 367
(c) 534
(d) 140
3. (a) four hundred fifty-five
(b) seven hundred forty-four
(c) eight hundred fifty
(d) nine hundred three
4. (a) >
(b) <
(c) >
(d) >
(e) >
(f) >

64

5. (a) 100 → **58** → **88** ↓ 70 → **99** → **59** ↑ 100

$$\begin{array}{ccccc} 100 & \rightarrow & \mathbf{58} & \rightarrow & \mathbf{88} \\ \uparrow & & & & \downarrow \\ \mathbf{59} & \leftarrow & \mathbf{99} & \leftarrow & 70 \end{array}$$

(b)

$$\begin{array}{ccccc} 1000 & \rightarrow & \mathbf{360} & \rightarrow & \mathbf{500} \\ \uparrow & & & & \downarrow \\ \mathbf{450} & \leftarrow & \mathbf{20} & \leftarrow & 320 \end{array}$$

65

6. (a) 287 + 195 = 482
482

(b) 287 – 195 = 92
92

(c) 482 + 170 = 652
652

66

3 Length

Objectives

- Compare the same lengths measured with different units.
- Estimate and measure lengths to the nearest meter or centimeter.
- Estimate and measure lengths to the nearest yard, foot, or inch.
- Solve word problems involving adding or subtracting lengths.
- Measure curved lengths.
- Use a ruler to draw lines of given lengths.
- Compare a meter to a yard.
- Compare a foot to a yard.
- Compare an inch to a centimeter.

Suggested number of weeks: 1-2

		TB: Textbook WB: Workbook	Objectives	Material	Appendix
3.1	**Measuring Length in Meters**				
3.1a	Meters	TB: pp. 52-53 WB p. 67	♦ Compare the measurement of two objects using different units. ♦ Estimate and measure length to the nearest meter.	♦ Large paper clips ♦ Small paper clips ♦ Meter sticks ♦ String, rope, or ribbon	
3.1b	Word Problems	TB: p. 54	♦ Add or subtract lengths in meters. ♦ Solve word problems.	♦ String, rope, or ribbon ♦ Meter sticks	a21
3.2	**Measuring Length in Centimeters**				
3.2a	Centimeters	TB: pp. 55-56 WB pp. 68-71	♦ Estimate and measure length to the nearest centimeter.	♦ Centimeter cubes ♦ Rulers ♦ Meter sticks ♦ Cardboard strips marked in centimeters ♦ String, rope, or ribbon ♦ Chart (see lesson)	
3.2b	Curved Lengths	TB: pp. 57-58 WB: pp. 72-73	♦ Measure curved lengths using string. ♦ Measure with a measuring tape. ♦ Use a ruler to draw lines of given lengths.	♦ String ♦ Measuring tapes ♦ Rulers ♦ Chart (see lesson)	a22

		TB: Textbook WB: Workbook	Objectives	Material	Appendix
3.3	**Measuring Length in Yards and Feet**				
3.3a	Yards and Feet	TB: pp. 59-60	♦ Estimate and measure length to the nearest yard or foot. ♦ Compare a meter to a yard. ♦ Compare a foot to a yard.	♦ Yard sticks ♦ Meter sticks ♦ String ♦ Rulers ♦ Foot-long cardboard strips	
3.4	**Measuring Length in Inches**				
3.4a	Inches	TB: pp. 61-62 WB: p. 74	♦ Estimate and measure length to the nearest inch. ♦ Compare an inch to a foot. ♦ Compare an inch to a centimeter.	♦ Rulers ♦ Charts (see lesson) ♦ Unmarked straight-edges ♦ Measuring tapes	
3.4b	Practice and Review	TB: p. 63 WB: pp. 75-78	♦ Practice addition and subtraction within 1000. ♦ Solve word problems involving measurement. ♦ Review all topics.		a23

3.1
3.2

Measuring Length in Meters or Centimeters

Objectives

- Compare the same lengths measured with different units.
- Estimate and measure lengths to the nearest meter.
- Add and subtract lengths in meters.
- Estimate and measure length to the nearest centimeter.
- Measure curved lengths.
- Use a ruler to draw lines of a given length.
- Solve word problems.

Material

- Large and small paper clips
- Meter sticks, rulers (1 ft/30 cm rulers)
- Cardboard strips 1 meter long.
- String, rope, or ribbon
- Unit cube from a base-10 set or centimeter cube
- Charts to record lengths (see lessons)
- Measuring tapes for each group of students
- Cardboard strips with centimeters marked
- Unmarked straightedge (such as a cardboard strip)
- Curved lengths worksheet (appendix)
- Chart for recording lengths of various body parts.
- Appendix p. a21-a22

Prerequisites

Students should understand the basic concept of length and understand the meaning of words pertaining to length, such as length, width, and height. In order to do the word problems, they should be able to add and subtract within 100.

Notes

In *Primary Mathematics* 1A, students learned to measure and compare length using nonstandard units, such as paper clips or blocks. In this unit, measuring in the standard units — meters, centimeters, yards, feet, and inches — are introduced.

Measuring with non-standard units in *Primary Mathematics* 1A allowed students to understand the nature of measurement. To measure the length of an object, we need to first define a unit object that has a measure of one. Once a unit of measure is defined, measurement is the process of determining how many copies of this unit fit, without overlap, along the length that we are measuring. The number that results from a measurement is dependent on the unit being used.

An object will have a different measure if the unit is different. Students should understand that a length measured with a shorter unit will result in a greater number of units than the same length measured with a longer unit. For example, the length of a pencil could be either about 5 long paper clips or about 9 short paper clips. A length of 9 is not necessarily longer than a length of 5 if the units used are not the same. Therefore, when giving the length of an object, is it essential to include the measurement unit.

Students should understand that when using a smaller unit, the numerical value of the measurement is larger. 9 short paper clips is the same as 5 longer paper clips. This may seem trivial, but understanding this concept thoroughly will help students later to remember what operation to use when they convert between measurement units in *Primary Mathematics* 3. When we convert from feet to inches, we multiply. 2 feet is 2 x 12 or 24 inches. Inches are the shorter unit so the number of inches will be greater than the number of feet. When we convert from inches to feet, we divide. 24 inches is 24 ÷ 12 or 2 feet. The number of feet will be smaller than the number of inches.

When measuring with either non-standard or standard units, every measurement is approximate. The accuracy of a measurement depends on the unit used. The smaller the unit, the more precise the measurement. For example, saying that something is about 1 meter long is less precise than saying that it is about 95 centimeters long. When measuring to the nearest unit of measurement, students can use the term *about*. The length of a pencil, for example, might be about 15 centimeters long.

Standard units of measurement are of vital importance in every civilization as they are an essential means of communication. At one time, sizes of body parts were used as units. However, since people don't share the same body sizes, it was not possible to communicate exact measurements since measured data depended on the measurer.

Students will be learning to measure in two distinct measurement systems, the metric system and the U.S. customary system. They need to keep the two systems separate. When they measure in compound units in *Primary Mathematics* 3B, they will be measuring length in meters and centimeters, or in yards, feet, and inches, but not in meters and inches.

The metric system has become the global language of measurement and today about 95% of the world's population uses the metric system. The use of the metric system is legal (but not mandatory) in the United States. The metric system is used in science.

The metric system of measurement is based on powers of 10. The prefixes tell what multiple of the basic unit is being considered. For length, the basic unit is a meter.

*Kilo*meter	1000 meters
*Hecto*meter	100 meters
*Deca*meter	10 meters
Meter	1 meter
*Deci*meter	0.1 meters
*Centi*meter	0.01 meters
*Milli*meter	0.001 meters

In Part 1 of this unit, the meter as a standard unit of measurement is introduced. In Part 2 of this unit, the centimeter as a standard unit of measurement is introduced. Students will be measuring in either meters or centimeters. In *Primary Mathematics* 3B, they will learn about millimeters. Hectometers, decameters, and decimeters are not taught in *Primary Mathematics*. Students will learn that 100 centimeters is the same as 1 meter, but will not be converting between centimeters and meters at this level.

Students should be able to estimate the length of an object in the various measurement units they are learning. To help them get a "feel" for the unit, they can use a familiar reference with which to compare the lengths in their minds. 1 meter is about the length of a baseball bat, or about the width of a door. They can relate their height to a meter. A dime is about 1 centimeter across. They can find a body part that has a measure of about a centimeter, such as the width of a thumb.

Most of the objects students measure will not be exactly a certain number of centimeters. Students learned about halves in *Primary Mathematics* 1B, so you can have them measure to the nearest half-centimeter or half-inch. Or you can have them just measure to the nearest whole unit for now. For example, if the length of an object is up to six and a half centimeters long, it is about 6 cm long.

In some of the practices or exercises, students will be asked to find the length around a figure. You can have students measure the length around a table top or a textbook without formally introducing the word perimeter.

3.1a Meters

Objectives

- Compare the measurement of two objects using different units.
- Estimate and measure length to the nearest meter.

Vocabulary

- Unit
- About
- Meter
- Abbreviation

Note

Meter sticks do not always start at 0 and end at 1 meter. In this lesson, students are estimating measurement to the nearest meter. It is not necessary to focus on measuring accurately from the 0 mark yet. Use string or cardboard strips that are 1 meter long so that students do not get distracted by the markings on a meter stick, or use the entire meter stick as a close enough approximation without. The purpose of these activities is to get students familiar with a meter and able to estimate length in meters, not to measure accurately, since most objects are not exactly a meter or a multiple of a meter long.

Review measurement with non-standard units	
Point to an object such as a table, and ask students how they would describe it to someone who has not seen it. The types of things we use to describe it, such as color, pattern on the fabric, size, etc., are attributes of the object. Get them to eventually talk about its length or width. In order to describe the length of an object to someone else, we have to measure it with something the other person is familiar with. We could say, for example, that the table is about the same length as a student.	
Provide students with paper clips of two different lengths. Have them measure a pencil or the length or width of their books with each type of paper clip. Point out that the length we are measuring does not always end at exactly the end of a unit. We therefore measure to the closest measuring *unit*. If a pencil is a bit longer than 5 paper clips, but not as long as 6 paper clips, we can say the pencil is *about* 5 paper clips long. Point out that the numerical values of the measurements are different depending on which type of paper clip is being used. So we always have to include what unit we are measuring with.	About 5 long paper clips. About 9 short paper clips.
Introduce meters	
Discuss the need for standard units of measurement. For example, have one student measure the length of a table with his or her hand spans. Then you measure the table with your hand spans. The numerical value of length is different. If we wanted to order a table ___ hand spans long, and used the student's hand spans, the table might be too long. So it is not possible to communicate exact measurements unless both people are using the same unit of measurement.	
Show students a meter stick. Explain that the *meter* is a standard unit of measurement. A meter everywhere is the same length. You could tell someone on the other side of the world that you wanted a table that was 2 meters long, and that person would know how long the table should be.	Meter sticks

Measure something in the classroom that measures about a meter with the meter stick, such as the width of a door. Tell students it is *about* 1 meter wide	
Write **1 meter = 1 m** on the board. Tell your student that **m** is the *abbreviation* for meter.	1 meter = 1 m
Determine if a length is less than or more than 1 meter long.	**Text pp. 52-53, Workbook p. 67**
Have students measure various items to determine whether they are longer than or shorter than a meter. You can do the suggested activities on page 52 and in Task 1 on page 53 of the textbook.	Answers will vary.
Have students do problems 1 and 2 for Exercise 25 in the workbook. Get them to note how high on their bodies 1 meter is so that they can think of that length as a reference point for a meter when making estimations for now. Or they can see how close to a meter their outstretched arms from hand to hand is. They can look for another reference they can easily visualize, such as the width of a doorway.	Meter sticks String, rope, or ribbon
As students measure objects, discuss the terms length and width. The word length means the measure from one end to the other, so all sides of the desk have a length. However, for a rectangular shape, we usually use the word *length* for the longer side, and *width* for the shorter side. The word *height* is usually used for the length going up and down.	Length Width Height
Estimate and measure length in meters	Text p. 53, Workbook p. 67
Have students measure lengths that are several meters long and say or record their answer as between ___ and ___ meters. You can use the suggested activity in Task 2 on page 53 of the textbook.	Meter sticks String, rope, or ribbon
Have students estimate lengths and record both their estimates and the actual lengths. You can have them fill out the chart in problem 3 of Exercise 25 in the workbook, or you can prepare a similar chart for groups of students to work on. Guide them in recording the lengths to the *nearest* meter.	
Assessment	
Point out some objects inside or outside of the classroom, or name some objects students are familiar with that are up to 5 meters in length and have students estimate their lengths.	

Exercise 25

Exercise should be done in class rather than as homework. Answers will vary.

67

3.1b Word Problems

Objectives

- Add or subtract lengths in meters.
- Solve word problems.

Note

This lesson is more a review of word problems than one in measuring lengths. It is a good way to review addition and subtraction from the previous lessons, and students will encounter many more word problems involving length later in reviews.

The pictures that accompany Tasks 4 and 5 in the textbook on p. 54 will lead to the part-whole and comparison models in *Primary Mathematics* 3. Do not require accurate drawings exactly like these yet.

Word problems	**Text p. 54**
Discuss the tasks on this page with the students, or have them work on the tasks individually or in groups and then discuss their solutions. Be sure they understand that the lengths given on the pages in the textbooks are not actual lengths, but have been scaled down, as in maps. The ribbons are pictures of ribbons that in "real life" are 7 m and 4 m long.	3. (a) 11 m (b) 3 m 4. 12 m 5. 36 m
Point out that the pictures help with solving the problems. So when they get problems without a picture, they can try to draw a picture and label the lengths to help them solve the problems.	
Assessment	Appendix p. a21
Have students write the equations needed. Encourage them to include the measurement unit (m for meters) with the numbers when they write the equations. A measurement without a unit is meaningless. Encourage them to draw simple diagrams if needed.	
1. Harriet needed to make some cages for her birds. She wanted to make two cages. For one, she needed 48 meters of chicken wire. For the other, she needed 76 meters of chicken wire. (a) How many meters did she need in all? (b) She bought 200 meters of wire. How many extra meters does she have?	48 m + 76 m = 124 m She needed 124 m. 200 m – 124 m = 76 m She has 76 m extra.
2. August lives 304 meters from school. May lives 283 meters from school. Who lives closer to the school? How much closer to the school does that person live?	May lives closer to school. 304 m – 283 m = 21 m She lives 21 m closer to school.
3. Three trees are in a line. The first tree is 105 m from the third tree. The second tree is 63 m from the third tree. How far is the second tree from the first tree?	105 m 63 m 105 m – 63 m = 42 m The second tree is 42 m from the first tree.

4. A skyscraper is 120 meters tall. A house is 104 meters shorter than the skyscraper. An apartment building is 38 meters taller than the house. (a) How tall is the house? (b) How tall is the apartment building? (c) If the house were put on top of the apartment, would it be as high as the skyscraper? How much higher or lower is the skyscraper than the apartment building and house together?	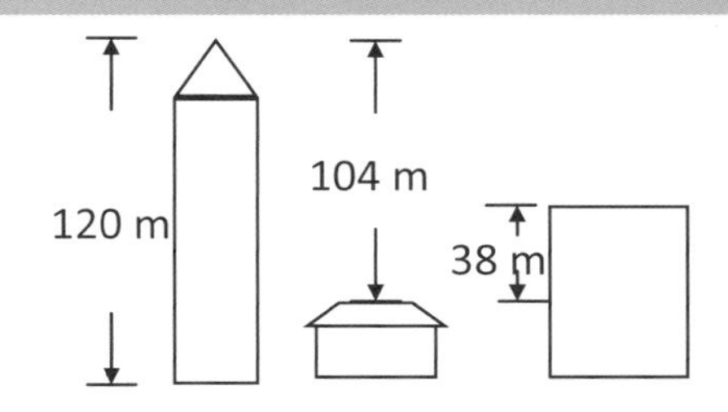 (a) 120 m – 104 m = 16 m The house is 16 m tall. (b) 16 m + 38 m = 54 m The apartment is 54 m tall. (c) 54 m + 16 m = 70 m 120 m – 70 m = 50 m The skyscraper is still taller by 50 m.
5. There are three ropes, A, B, and C. The total length of the ropes is 94 meters. Rope A is 51 meters long. Rope C is 38 meters long. How long is Rope B? This is a 2-step problem. Students can either subtract the length of Rope A and then the length of Rope C from the total, or first add the lengths of Ropes A and C together and subtract that from the total length. Encourage them to draw some pictures of the ropes, or number bonds. Note that it is not necessary that the ropes in the pictures are proportional. Students will not know ahead of time that Rope B is so short compared to the others.	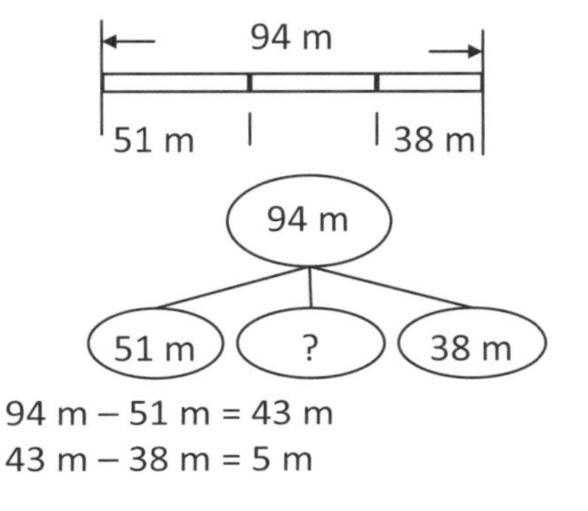 94 m – 51 m = 43 m 43 m – 38 m = 5 m 51 m + 38 m = 89 m 94 m – 89 m = 5 m Rope B is 5 m long.

Activity

Material: Two lengths of string, rope, or ribbon for each group of students. They should be longer than 10 meters, long enough so that it is difficult to measure just by laying them out straight, and measure very close to the meter. Meter sticks or cardboard strips 1 meter long.

Procedure: Give each group two of the strings, ropes, or ribbons and a meter stick and ask them to find the total length of both and how much longer one is than the other. They will need to come up with a way of measuring them (such as measuring a part at a time, wrapping them around the meter stick, or cutting them into 1 meter lengths).

3.2a Centimeters

Objectives

- Estimate and measure length to the nearest centimeter.

Vocabulary

- Centimeter

Note

In order to allow students to concentrate on centimeters and not the other marks on a ruler initially, you may want to use cardboard strips marked in centimeters only. Later in the lesson, you can distribute rulers and have them compare both sides to the cardboard strip to see which side would be used to measure in centimeters, and which mark is for the centimeter. With the rulers, be sure that students are measuring using the side marked in centimeters. Make sure they measure from the 0 mark on a ruler, which, depending on the ruler, may not be at the edge of the ruler.

Introduce centimeters	
Ask students if they noticed any difficulties with measuring length to the nearest meter. We would still not be able to easily tell someone what length of table we wanted if it was between 1 and 2 meters. We need a standard unit of measurement for objects that are shorter than a meter. Tell them that one shorter unit of measurement that is standard all over the world and used a lot is called a *centimeter*.	
Show students a unit cube from a base-10 set or a centimeter cube and tell them it is about a centimeter on each side. Have students look at their rulers or cardboard strips. If you are having them use rulers, be sure they know which side is for centimeters and which marks they should use. You can tell them the shorter marks are for even smaller units to measure even smaller lengths, but that we will only be measuring to the nearest centimeter for now.	Centimeter cube Rulers or cardboard strips
Have students examine the meter stick and the marks on it. Guide them in knowing which marks show centimeters. Ask them to determine how many centimeters are in a meter. There are 100 centimeters in a meter.	Meter sticks
Write **1 centimeter = 1 cm** on the board and tell students that **cm** is the abbreviation for centimeter.	1 centimeter = 1 cm
Measure length in centimeters	**Text pp. 55-56**
Have students look at p. 55 in the textbook. They can compare their rulers with the ruler shown on that page. Ask them to measure the picture of the grasshopper and the fish with their own rulers. Guide them in locating the 0 mark and measuring from that. Point out the abbreviation for centimeters (cm). Have students do the tasks on p. 56 of the textbook. Get students to find some part of their hands that measures 1 cm closely, such as the width of one of their fingers. This will give them a reference that they can use for estimation.	The grasshopper is **2** cm shorter than the fish, and the fish is **2** cm longer than the grasshopper. 1-2. Answers will vary. The *Primary Mathematics* textbook is about 26 cm long and 19 cm wide.

Estimate and measure length in centimeters	**Workbook p. 68**
Have students first estimate and then measure the lengths of some objects. You can prepare a chart like that in problem 1 of Exercise 26, using items in the classroom. Use lengths that are less than 30 centimeters. Save the chart to be used with inches.	1. Answers will vary.
Have students measure some lengths that are greater than 30 centimeters, using a single ruler. They need to measure 30 centimeters, mark that, and then put the start of the ruler at that mark and continue measuring. They must then add to find the total length.	
Ask students to measure the length around a flat object with straight sides, such as a textbook or a table top. Get them to measure each side and then find the total distance around the object by adding the length of each side.	Distance around the *Primary Mathematics* textbook: 26 cm + 19 cm + 26 cm + 19 cm = 90 cm
Assessment	
Point out certain items, such as a tree or book, and ask students if we would measure it in meters or centimeters. Ask them to estimate the lengths.	
Do a length hunt. Give students a length in centimeters and have them find something that is close to that length and then check with their rulers. They can use a meter stick for lengths longer than 30 cm.	
Give each group of students two lengths of string or ribbon that are each under 100 cm and ask them to find the total length and how much longer one is than the other.	String or ribbon
Practice	WB Exercise 26, pp. 68-71, problems 2-5

Exercise 26

1. Answers will vary.

2. (a) 3 cm
   (b) 10 cm

68

3. (a) 8 cm
   (b) 11 cm
   (c) 9 cm
   (d) 12 cm

69

4. (a) 11 cm
   (b) 9 cm
   (c) 2 cm
   (d) 4 cm
   (e) brush
   (f) clothes pin

70

5. (a) m
   (b) m
   (c) cm
   (d) cm
   (e) m
   (f) cm
   (g) cm
   (h) cm
   (i) m
   (j) cm

71

3.2b Curved Lengths

Objectives

- Measure curved lengths using string.
- Measure with a measuring tape.
- Use a ruler to draw lines of given lengths.

Note

Measuring a curved length with string takes a good deal of coordination, so you may have to demonstrate on the board or help individual students. They need to put the start of the string at the start of the curved line, and then match it up with the line by laying it along the line and holding it down with their fingers, moving fingers along the string. The end can be marked with a marker or cut.

Drawing straight lines using a ruler can also be tricky. Guide students in holding the ruler firmly enough so that they can move the pencil along the edge without dislodging the ruler.

Measure curved lengths	
Find some item in the classroom that has a curved edge. Ask students how we could measure the length of that edge — the length it would be if it could be pulled straight. You may want to let them try out their suggestions. One method is to lay some string along the edge, cut the string or mark it, and then measure the length of the string with a ruler. Draw a curved line on the board and demonstrate how we can measure it with a string.	String
Measure curved lengths	**Text p. 57**
	3. 14 cm 4. Line A is about **12** cm long. Line B is about **12** cm long. They are the same length. 5. Papaya Road is shortest. Rambutan Road is longest.
Measure with a measuring tape	**Text p. 58, Workbook p. 72**
Have students work in groups. Give each group a measuring tape to examine. Have them do Task 6 on p. 58 of the textbook, using the chart for problem 1 of Exercise 27, or you can create a similar chart as a handout. Guide them in using the measuring tape properly. You can also provide each student with a list of body parts and let them measure different parts of their bodies — wrist, arm length, circumference of head, length of foot, etc. — and compare the lengths with each other. For example, is the length of a foot the same as the distance around an arm?	6. Answers will vary. Measuring tapes

Draw straight lines with a ruler	**Text p. 58**
Guide students in using a ruler to draw lines.	7. Check lines.
Have students draw additional lines of specified lengths in centimeters.	Rulers
Assessment	Appendix p. a22
Give students copies of appendix p. a22 and have them measure the curved lengths to determine which is longest.	
Ask students to draw a straight line 12 cm long.	
Practice	WB Exercise 27, pp. 72-73, problems 2-4

Activity

Material: Unmarked straightedge for each student, rulers.

Procedure: Divide students into groups. One student calls out a length that is less than 30 cm. The rest use the straightedge to draw a line they think is that length, and then measure to see who came closest to the desired length. Students take turns calling out a length to draw.

Exercise 27

1. Answers will vary.
2. (a) 12 cm
   (b) 9 cm
   (c) 3 cm

72

3. Length: about 12 cm
   Width: about 4 cm
4. (a) 8 cm
   (b) 11 cm
   (c) 10 cm
   (d) B
   (e) A

73

3.3 3.4 Measuring Length in Yards, Feet, and Inches

Objectives

- Estimate and measure length to the nearest yard, foot, or inch.
- Compare a meter to a yard.
- Compare a foot to a yard.
- Compare an inch to a foot.
- Compare an inch to a centimeter.
- Solve word problems involving length.

Material

- Yard sticks
- Meter sticks
- Rulers
- Measuring tapes
- Cardboard strips or string 1 foot long for each student
- Unmarked straightedges
- Charts to record measured and estimated lengths (see lessons)
- Appendix p. a23

Notes

In Part 3, students will be introduced to yards and feet as units of measurement. In Part 4, they will be introduced to inches.

Make sure students understand that they are learning to use two different systems of measurement. One is the metric system, which is used in most of the world and in science. It includes meters and centimeters. The other is sometimes called the U.S. customary system and includes yards, feet, and inches.

Students should be able to estimate lengths. Estimation is easier when they have an easily remembered reference. A yard is close enough in length to a meter that students can use the same estimate for that, such as the distance from hand to hand if the arms are outstretched. The length of a forearm might be about a foot. Students will most often be measuring in inches or centimeters and should be able to make reasonable estimates for the length of objects up to 12 inches. The length of the last joint of a finger might be an inch. The length of a hand span might be about 6 inches.

Students will not be converting between different measurement systems in *Primary Mathematics*, but they should be able to understand the comparative sizes of the measurements. A meter is just a little longer than a yard. If something is exactly 6 yards long, it will be less than 6 meters long. A centimeter is shorter than an inch. Students should realize that the numerical value of a measurement in centimeters will be greater than the same measurement in inches. From the ruler, they can see that 30 centimeters is about 12 inches.

Miles and kilometers will be taught in *Primary Mathematics* 3B, but they can be mentioned here if the subject comes up. A kilometer is a little more than half of a mile.

The abbreviation for yard is **yd** and for foot or feet it is **ft**. The only measurement where a period is used in the abbreviation is the inch (**in.**) to distinguish it from the word *in*. (Conventions change, and some spell checkers will want periods on other measurement abbreviations than inches.)

Since a foot-long ruler is not exactly 1 foot, use cardboard strips that are exactly 1 foot for activities where they need to measure to the nearest foot, or use string as suggested in the text. However, the difference between a foot and the length of a ruler is small enough that the length of a ruler is adequate when measuring to the nearest foot. Students will mostly be measuring objects to see if they are longer or shorter than a foot or a meter, rather than trying to get a precise measurement. You can discuss with students when measurements need to be precise and when an approximate measurement is sufficient.

In Part 4, students will learn that the 12 inches on the ruler is 1 foot, so you may want wait until then to use actual rulers rather than cardboard strips a foot long.

It is the metric system, not the U.S. customary system, that is used whenever accuracy and precision are important. Almost all the technical measurements used in hospitals, industry, and science in the U.S. use the metric system. The U.S. army uses the metric system exclusively. In this curriculum, more time is spent on the metric system than on the U.S. customary system, and it is taught first along with other concepts involving measurement.

There is no customary unit of measurement in the U.S. smaller than an inch. In *Primary Mathematics* 3 students will learn about the millimeter. On rulers, centimeters are divided into tenths. Each tenth is a millimeter. Inches, on the other hand, are generally divided into halves, fourths, eighths, and sixteenths with different lengths for the markings. Since students have learned about halves and fourths, you can get them to measure to the nearest half. Be sure they can interpret the scale on the ruler. They will have more experience with scales and fractions later, so measuring to a fraction of an inch can wait.

3.3a Yards and Feet

Objectives

- Estimate and measure length to the nearest yard or foot.
- Compare a meter to a yard.
- Compare a foot to a yard.

Vocabulary

- Yard
- Foot
- Feet

Note

Cardboard strips exactly a foot long will be easier to use to measure straight lengths with than string. When students compare a foot to a yard to find that there are 3 feet in a yard, they need to be fairly accurate, so they will need cardboard strips or strings that are exactly a foot long and a meter long.

You can have more capable students measure to the nearest half-foot or quarter-foot by estimating where halves or quarters are on the measuring device, or you can mark fourths on the rulers if they are using cardboard strips.

Introduce yards	
Tell students that in the U.S. we often measure length in a unit called the *yard*. Show students a yard stick, and point out the mark that shows them a full yard. Have them compare that to the mark that shows a full meter on a meter stick or on the other side of the yard stick if it is marked with both systems. A meter is a little longer than a yard.	Yard sticks Meter sticks
Write **1 yard = 1 yd** on the board and tell students that **yd** is the abbreviation for a yard	1 yard = 1 yd
Estimate and measure length in yards	**Text p. 59**
You can have students do any of the activities that seem appropriate on p. 59 of the text and estimate lengths that are longer than a yard to the nearest yard.	1. Answers will vary. 2. Answers will vary.
Have students find a body part to use as reference for a yard and a foot in order to have some length in mind when making estimates. They can use the same reference they found for a meter, since a yard is about a meter.	
Introduce feet	
Tell students that another unit of measurement is the *foot*. Show them a foot ruler or a foot-long cardboard strip. Tell them that we say 1 *foot*, but 2 or more *feet*.	Rulers or cardboard strips
Write **1 foot = 1 ft** and **2 feet = 2 ft** on the board. Tell them ft is the abbreviation for both foot and feet.	1 foot = 1 ft 2 feet = 2 ft
Estimate and measure length in feet	**Text p. 60**
Provide students with foot-long cardboard strips or string and have them measure various lengths to the nearest foot.	3. Answers will vary. 4. 1 yard = **3** feet
Get students to find a body part that is close to 1 foot, such as their forearm or 2 hand spans.	

You can tell students that the length and width of rooms in a house are often given in feet. Have them measure the length and width of the classroom to the nearest foot. Give your height in feet or have them use their height in feet and get them to estimate how high the ceiling is. For homework, you might want to ask them to measure the widths and lengths of some of the rooms in their houses.	
Assessment	
Point to various objects in the classroom or outside the window and ask whether we would measure them in yards or feet. A flagpole, for example, would usually be measured in yards. The distance around the playground or field track would be measured in yards. The height of a room or length of a ladder would be measured in feet. See if they can give a reasonable estimate for the number of yards or feet.	

3.4a Inches

Objectives

- Estimate and measure length to the nearest inch.
- Compare an inch to a foot.
- Compare an inch to a centimeter.

Vocabulary

- Inch

Note

You can have students measure to the nearest half-inch and quarter-inch now, or wait until *Primary Mathematics* 3B after they have had more experience with fractions and scales. If you do have them measure to the nearest half-inch, guide them in determining which marks on the ruler are for a half of an inch.

The ruler on the top of p. 61 in the textbook is not accurate in the current printing. Students cannot compare their rulers to it.

Introduce inches	**Text p. 61**
Tell students that in the U.S. we often measure lengths shorter than a foot in *inches*. Guide students in locating the inch marks on their rulers. Then have students look at p. 61 in the text. Tell them that the ruler and pictures at the top are scaled down slightly and are not exactly an inch, so they should not measure the picture of the paper clip and the pen with their rulers. You can have them answer the questions on this page, or measure two other items in actual inches and tell you how much shorter or longer one item is than the other.	Rulers
Tell students that 12 inches is a foot. Write **1 foot = 12 inches** and **1 ft = 12 in**. on the board. Tell them that **in.** is the abbreviation for inch. For this abbreviation we put a period at the end to show that it is an abbreviation and not the word *in*.	1 foot = 12 inches 1 ft = 12 in.
Have students compare the inch to a centimeter using their rulers. Have them look at p. 62 in the textbook.	**Text p. 62**
Measure to the nearest inch	**Text p. 61**
Have students measure the length and width of their textbooks to the nearest inch and compare that to the same measurement in centimeters.	1. Answers will vary. The *Primary Mathematics* textbook is about 10 inches long and 8 inches (or 7½ inches) wide.
Have students determine which joint of which finger is closest to an inch. You may want to share with your students a bit of trivia: The term *rule of thumb* means an easily learned and easily applied procedure for approximately calculating or recalling some value, or for making some determination. The term is thought to originate with woodworkers who used the length of their thumbs rather than rulers for measuring things. In an adult, the last joint in the thumb is about an inch long, with the thumbnail being about a half-inch long.	

Have students find the measurement in inches all the way around an object with straight sides, such as a desktop. They can find the length of each edge and then add them, or add as they go.	
Estimate and measure length in inches	
Have students estimate some lengths in inches and then measure. They can use the same charts as they used in lesson 3.2a and compare the measurements in inches to those in centimeters.	
Have students use a ruler to draw lines of specified lengths in inches.	Rulers
Have students use an unmarked straightedge, such as the cardboard rulers, draw a line that they think is a given number of inches, and then measure with a ruler to see how close they came.	Unmarked straightedges
You can have students use measuring tapes to measure various body parts in inches. Make sure they know which side of the tape has inches. They can compare these to the same measurements in centimeters.	Measuring tapes
Assessment	
Point to several items and tell students the numerical length. Get them to tell you what unit must have been used (yards, feet, or inches). For example, the length of a book is 10 what? The height of the ceiling is 10 what? The width of the hallway is 10 what?	
Tell students that a bridge is 6 yards long. Ask them if it is more than or less than 6 feet long. Ask them if it is more than or less than 6 meters long.	6 yd > 6 ft 6 yd < 6 m
Tell students that you have a pencil that is 6 inches long. Is it more than or less than 6 centimeters long?	6 in. > 6 cm
Practice	WB Exercise 28, p. 74

Exercise 28

1. (a) ft
   (b) in.
   (c) yd
   (d) in.
   (e) ft
2. Yellow rod
3. Check line drawn.
   It should be 6 inches long.
4. Check line drawn.
   It should be 1 inch long.

74

3.4b Practice and Review

Objectives

- Practice addition and subtraction within 1000.
- Solve word problems involving measurement.
- Review all topics.

Note

At this level, word problems with measurements simply involve adding or subtracting lengths. Students can solve them in the same way they have been solving word problems in previous units. If they have any difficulties with solving the problems, they can draw number bonds, or draw pictures that look more like the objects in the problem, such as a line for a piece of ribbon or a stick figure or vertical line for height, and indicate what they want to find with a question mark.

Practice	Text p. 63, Practice 3A
Students can do these problems individually and then share and discuss their solutions to the word problems. Some possible solutions for problems 7 and 9 are provided below as suggestions. Problem 7: The whole is the length of the ribbon, 90 cm, and one part is cut off. To find the part used for a bow, we subtract. Ribbon 90 cm Left 35 cm Bow ? Ribbon: 90 cm 35 cm Used for bow Problem 9: The whole is Taylor's height. One part is how much shorter Nicole is, and the other Nicole's height. This is a comparison problem. Taylor 96 cm Nicole ? Shorter 8 cm Taylor 96 cm 8 cm Nicole ?	1. (a) 294 (b) 399 (c) 500 2. (a) 571 (b) 502 (c) 960 3. (a) 384 (b) 187 (c) 378 4. (a) 129 (b) 204 (c) 319 5. (a) 800 (b) 178 (c) 694 6. 350 m + 550 m = **900 m** Samy walked 900 m. 7. 90 cm – 35 cm = **55 cm** She used 55 cm for the bow. 8. 24 yd + 12 yd + 12 yd + 16 yd = **64 yd** The total length around the field is 64 yd. 9. 96 cm – 8 cm = **88 cm** Nicole's height is 88 cm.

Enrichment	Appendix p. a23
1. A snail is crawling up one of the posts Harriet made for her birds to perch on. The post is 39 inches tall and is slippery. If the snail crawls up the post 15 inches every hour, but slides back 7 inches when he rests at the end of each hour, how many hours will it take the snail to get to the top? If we solve this by assuming the snail crawls up 8 inches an hour, he would reach the top during the fifth hour. However, the problem states that he rests at the end of the hour. So the climb only takes 4 hours, since he is within 15 inches of the top after his rest at the end of the third hour.	Each hour, it would move forward 8 inches, so 1 hr $\longrightarrow$ 8 inches 2 hr $\longrightarrow$ 8 + 8 = 16 inches 3 hr $\longrightarrow$ 16 + 8 = 24 inches 4 hr $\longrightarrow$ 24 + 8 = 32 inches 5 hr $\longrightarrow$ 32 + 8 = 40 inches But, 24 inches + 15 inches = 39 inches It takes him 4 hours.
2. There are 4 sticks. Stick D is between Stick A and B. Stick C is the shortest. Stick B is shorter than Stick A. Arrange the sticks in order, starting with the shortest stick.	D is between A and B: ADB or BDA C is the shortest: CADB or CBDA B is shorter than A: CBDA CBDA
Review	WB Review 2, pp. 75-78

Review 2

1. (a) 99
   (b) **8** tens **2** ones
   (c) **6** hundreds **4** tens **7** ones
   (d) **5** hundreds **0** tens **3** ones

2. 899, 904, 908, 910

3.

994	**995**	996	997	**998**	**999**
	985				
	975		977	**978**	**979**
	965			968	
	955	956	**957**	**958**	
	945			948	

75

4. (a) >
   (b) <
   (c) >
   (d) >
   (e) <
   (f) >

5. $223

6. (a) 157 (b) 873
   (c) 209 (d) 920

7. (a) 30 (b) 42
   (c) 8 (d) 100

76

8. (a) 1000
   (b) 800
   (c) 690
   (d) 308
   (e) 242

9. (a) 349
   (b) 748
   (c) 604
   (d) 580

9. (a) two hundred twenty
   (b) four hundred thirty-one
   (c) eight hundred sixty-nine
   (d) nine hundred forty-four

77

8. (a) $45 – $26 = $**19**
   (b) $380 + $18 = $**398**
   (c) $26 + $6 + $8 = $**40**

78

4 Weight

Objectives

- Estimate and weigh objects in kilograms or grams.
- Estimate and weigh objects in pounds or ounces.
- Use and read weighing scales.
- Compare a pound to a kilogram.
- Compare an ounce to a gram.

Suggested number of weeks: 1-2

		TB: Textbook WB: Workbook	Objectives	Material	Appendix
4.1	**Measuring Weight in Kilograms**				
4.1a	Kilograms	TB: pp. 64-65	♦ Estimate and weigh objects to the nearest kilogram using a balance.	♦ Balance ♦ Kilogram weights ♦ Items to weigh ♦ Beans ♦ Resealable bags ♦ Chart (see lesson)	a24
4.1b	Weighing Scales	TB: pp. 65-67 WB: pp. 79-80	♦ Weigh objects to the nearest kilogram using scales. ♦ Read scales.	♦ Weighing scale ♦ Chart (see lesson)	
4.2	**Measuring Weight in Grams**				
4.2a	Grams	TB: pp. 68-69 WB: pp. 81-82	♦ Estimate and weigh objects in grams.	♦ Paper clips ♦ Unit cubes ♦ Balance ♦ Chart (see lesson)	a25-a26
4.3	**Measuring Weight in Pounds**				
4.4	**Measuring Weight in Ounces**				
4.4a	Pounds and Ounces	TB: pp. 70-73	♦ Estimate and measure weight in pounds. ♦ Estimate and measure weight in ounces. ♦ Compare a pound to a kilogram. ♦ Compare an ounce to a gram.	♦ Balance ♦ Pound and ounce weights ♦ Scale marked in pounds and ounces ♦ Bag of 58 paper clips ♦ Charts (see lesson)	a27
4.4b	Practice and Review	TB: pp. 74-75 WB: pp. 83-90	♦ Practice addition and subtraction within 1000. ♦ Solve word problems involving measurement. ♦ Review all topics.		

Blank Page

4.1 4.2 Measuring Weight in Kilograms and Grams

Objectives

- Estimate and weigh objects to the nearest kilogram using a balance.
- Weigh objects to the nearest kilogram using scales.
- Read scales.
- Estimate and weigh objects in grams.

Material

- Balances
- Uniform objects to use as non-standard weights, e.g., blocks
- Kilogram and gram weights
- Resealable plastic bags
- Dried beans
- Various objects weighing less than 5 kg
- Various measuring scales, such as a bathroom scale
- Various objects labeled with their weight in grams
- Measuring scale for grams
- Paper clips or unit cubes
- Charts (see lessons)
- Appendix pp. a24-26

Prerequisites

Students should understand the basic concept of weight, and understand the meaning of words pertaining to weight, such as heavy, light, and weigh. In order to do the word problems, they should be able to add and subtract within 100.

Notes

In *Primary Mathematics* 1 students learned to compare weights by feel or *heft* as well as with a simple balance. They also measured weight using non-standard units.

In Part 1 of this unit students will estimate and weigh objects in kilograms. In Part 2 they will estimate and weigh objects in grams.

A liter of water weighs 1 kg. There are 1000 grams in a kilogram.

The kilogram and the gram are a measure of mass, not weight. We often use the terms *mass* and *weight* interchangeably in our daily speech, but to a scientist they are different things. The mass of a body depends on how much matter it contains. Mass is a measure of how much inertia an object displays, that is, how much force is needed to get it moving from a resting position or to stop it once it is moving. Weight, in contrast, is a measure of the gravitational pull between two objects. A 1 kg mass on Earth would weigh less on the moon but still have the same mass. Mass is measured by using a balance comparing a known mass to an unknown mass. Weight is measured on a scale where the gravitational pull is calibrated. Scales which measure in kilograms or grams are calibrated only for the equivalent mass on Earth. The metric unit for weight is the *Newton*; on earth a mass of 1 kilogram weighs 9.8 Newtons.

The pound and the ounce, units of weight in the U.S. customary system, which will be introduced later, are measures of weight, not mass. The slug is the English measurement of mass, and on earth a mass of one slug weighs 32.2 pounds.

Students will be weighing things only up to about 5 kilograms, primarily to see if they are more or less than You will not need kilogram weights for each student since they will be making kilogram weights with bags of beans. Also, if you have one item that weighs 1 kilogram, you can use it to make additional weights out of non-drying modeling clay. Clay often comes in weight amounts and can be used to make an amount close to 1 kg. For example, a pound of modeling clay often comes in 4 sticks, each a quarter of a pound. 9 sticks are very close to 1 kilogram.

Before the balance is used, make sure it is centered; that is, the fulcrum is placed so that one side is balanced with the other side before adding any weights. Simple classroom balances have a slider for centering the balance.

Students will also be weighing objects in grams. A gram is very light; the unit cube of a base-10 set or two small paper clips weigh about a gram. A simple primary balance will not weigh accurately to a gram, but can probably weigh to an accuracy of 50, and possibly 20 grams.

If enough weights are not available for students to work in groups, you can make weights with other objects. A penny weighs about two and a half grams, so you can make 10 gram weights by taping 4 pennies together and 50 gram weights by taping 20 pennies together. Or you can see how many counters or multilink cubes weigh about 50 grams. At this stage students do not have to be very accurate in their measurements, so exact weights are not necessary. Provide a variety of weights — e.g., 50-gram, 100-gram, and 500-gram weights — so that students can experiment with combinations to balance whatever object they are weighing.

Students will be reading the scale on weighing scales and need to determine what each mark on the scale means in the pictures in the textbook. Even if you don't have a scale for each group of students, have one you can demonstrate with and they can experiment with. Some scales show both kilograms and pounds, with divisions for grams and ounces. They won't look like the pictures in the book, and the scales will likely be different. Provide students with as much help as needed in reading the scales rather than focus on teaching them to read the scales. Reading scaled number lines will be covered again in *Primary Mathematics* 2B in the context of bar graphs and does not have to be mastered yet.

Students should gain some idea of how heavy a kilogram and a gram are. They can associate the weight with something whose weight they are familiar with. For example, two soccer balls, ten medium-sized apples, or two cans of soup weight about a kilogram. Since a gram does not have much weight (you can hardly feel the weight of 2 paper clips) they should know how much some familiar objects weigh in grams. For example, a medium-sized apple weighs about 100 grams.

How long each lesson takes depends on what resources you have (balances, weights, scales), how many objects students weigh, and how large your groups are. Adjust the lessons as needed.

Although many weighing scales are electronic, even bathroom scales, using balances and mechanical or spring scales where students see the needle or the scale move will give students a better understanding of the concept of weight and how scales work than simply seeing the numbers appear.

4.1a Kilograms

Objectives

- Estimate and weigh objects to the nearest kilogram using a balance.

Vocabulary

- Weight
- Weigh
- Kilogram

Note

The emphasis in this lesson is to give students an idea of how heavy a kilogram is, not to weigh accurately. They should be able to estimate whether something weighs more than, less than, or about 1 kilogram.

Weight	
Remind students they have been looking at one attribute or way of describing an object, length. Tell them that another attribute of an object is how heavy it is, that is, its *weight*. Discuss various ways in which weighing is used in everyday life. Have students tell you some things that we *weigh* and give you some reasons why we might want to weigh things. Discuss ways in which we could compare the weight of two objects. A seesaw is often used as an example of a way to compare weights, though both people have to be sitting the same distance from the fulcrum.	
Show students a balance, two objects to weigh, and some uniform blocks, such as multilink cubes and counters. Ask students for suggestions on how to compare the weights of the two objects. Obviously we can compare the weights directly by putting each on either side of the balance. Ask what we could do if we did not have both objects available at the same time. We could find out how many cubes each weighs and compare the number of cubes. You can demonstrate or have students work in groups to find the weight of two objects in nonstandard units. Tell students that if we use a different unit, such as counters, the object would have a different weight. You can weigh one of the objects against counters and discuss why more counters are needed than cubes.	Balance 2 objects Non-standard weights
Introduce the kilogram	
Tell students that, as with length, we use standard units to measure weight. Show them a kilogram weight and let them pass it around to feel its heft. Tell students that a *kilogram* is a standard unit of weight and is the same everywhere in the world.	Kilogram weight
Write **1 kilogram = 1 kg** on the board. Tell students that **kg** is the abbreviation for kilogram.	1 kilogram = 1 kg

Make kilogram weights	Text pp. 64-65
Have students work in groups. Provide each group with a balance, a kilogram weight, some resealable plastic bags, and beans. Guide them in creating bags of beans that weigh 1 kg (Task 1). Some balances have a slider that lets the balance be zeroed (balanced without weights). Show students how to zero such balances before using them.	Balances Kilogram weight Beans Resealable bags
Estimate and weigh	
Have students compare the weight of various objects to 1 kilogram. They should first "weigh" the object in their hands and decide whether it weighs more or less than a kilogram, and then check with the balance.	Balances Kilogram weight
You can have students weigh some objects that weigh more than 1 kg using their kilogram weights. They will have to put the object on one side and then add kilogram weights, such as the bags of beans, to the other until the weight tips. They can give the weight as "between 3 and 4 kilograms," for example. They can approximate which weight the object is closest to by how much the balance tips. Have them record the weights.	Object | Estimate | Weight (blank table, 3 rows)
Assessment	
Point to various objects in the room and ask students if they think the objects weigh more or less than a kilogram. Where appropriate, allow them to lift the object. You can ask them to estimate the weight of some objects to see of they can give reasonable estimates. Ask students which weighs more, a kilogram of feathers or a kilogram of rocks?	Which weighs more, a kilogram of feathers or a kilogram of rocks? They weigh the same, 1 kg.
Enrichment	Appendix p. a24
1. Box A is heavier than Box B. Box B weighs the same as Box C. Box C is heavier than Box D. Which is lightest, A, B, C, or D? Since B is lighter than A, C must also be lighter than A. D is therefore lighter than A and B as well.	A B B C C D D is the lightest.
2. Box X and 2 kg weigh the same as Box Y and 3 kg. How much heavier is Box X than Box Y? The sides would stay balanced if 2 kg were removed from both sides. Box X must be 1 kg heavier than Box Y.	X 1 kg 1 kg Y 1 kg 1 kg 1 kg Box X is 1 kg heavier than Box Y.

4.1b Weighing Scales

Objectives

- Weigh objects to the nearest kilogram using scales.
- Read scales.

Vocabulary

- Scale

Note

Students have used number lines before, but mostly with a scale of 1. Here and in the next few lessons they will be reading curved number lines where not all divisions of 1 are marked. Students need to be aware of the labeled marks on the scale, and the divisions between those marks. For students who have difficulty, it is not necessary that they master the concept of reading scales at this time as long as they can read the weighing scales used in class and in the textbook. They may find scales easier to understand after they have more experience with bar graphs, where they will need to read the scales on the axes of the graphs.

Add and subtract weights

Use the results from the activity in the previous lesson where students weighed objects to the nearest kilogram. List the objects and their weights to the nearest kilogram. Or list the weights of other objects in order to use weights more than 5 kilograms. Some examples are shown at the right.
Ask students to order the objects from lightest to heaviest.
Ask students questions such as:
Which object weighs the least?
Which objects weighs the most?
How much more does the lion weigh than the horse?
How much do they weigh altogether?
You can have students write appropriate equations.

Object	Weight
Horse	113 kg
Large Dog	27 kg
Cat	4 kg
Lion	155 kg
Cow	210 kg

Weigh using a scale

WB Exercise 29, problems 1-2

Show students a weighing scale. Have them observe how the needle moves around as you press down on it. Point out that we cannot use the scale to weigh anything heavier than the greatest weight marked on the scale.

Have students estimate and weigh some items on the scale to determine if they are more or less than 1, 2, or 3 kilograms, depending on the maximum weight for your scale.

If you have more than one scale, students can do this in groups. Guide them in interpreting the marks and divisions on the scale. You can have them do problems 1 and 2 from Exercise 29 in the workbook, or provide them with other charts and objects or groups of objects to weigh.

Weighing scale

Object	Estimate	Weight

Give students some uniform objects, such as blocks, and have them find out how many come closest to weighing 1 kg. Or, you can give them some clay and see if they can get a lump of clay that weighs exactly 1 kg, using the scale.

Interpret and compare scales

If you have two different weighing scales, you can discuss the marks on both and weigh the same item on both to show how far around the needle goes for each.

Or, draw two number lines with different scales, such as those shown at the right. Point out the distance from 0 to 1 on each scale. Point out that a number, such as 3, is at a different place depending on the number line being used and what the scale, or the distance between 0 and 1, is.

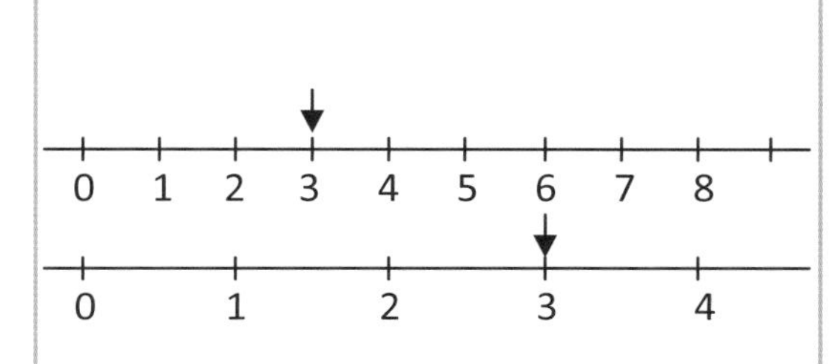

Read scales

Text p. 67

Have students look at the scales in Task 6 at the top of p. 67 in the textbook. Ask what the maximum weight each of these can be used for is. The one on the left can weigh objects up to 6 kg, and the one on the right can weigh objects up to 10 kg. Point out that the scales are different. It takes more weight to push the needle around on the scale on the right. Ask students to answer the questions. Make sure they understand why the package on the right weighs more, even though the needle moves around farther for the package on the left. The scales are different; the one on the right needs more weight to move the needle.

6. (a) The green package weighs 4 kg and the purple one weighs 6 kg. The **purple** one is heavier by **2 kg**.
   (b) 4 kg + 6 kg = **10 kg**
   The total weight is 10 kg.

Assessment

Text pp. 65-67

2. (a) 2 kg
   (b) 3 kg
3. Lighter than 1 kg
4. More than 2 kg
5. 9 kg
7. (a) Blue weighs 1 kg. Orange weighs 4 kg. Green weighs 5 kg. **Green** is heaviest.
   (b) **Blue** is the lightest.
   (c) 1 kg + 4 kg + 5 kg = **10 kg**
   The total weight is 10 kg.

Practice

WB Exercise 29, pp. 79-80, problems 3-4

Exercise 29

1. Do in class. Answers will vary.
2. Do in class. Answers will vary.
3. The **papaya** (**2nd** fruit) weighs more than 1 kg.
   The **grapes** (**3rd** fruit) weigh 1 kg.
   The **bananas** (**1st** fruit) weigh less than 1 kg.

79

4. (a) 2 kg (b) 3 kg
   (c) 7 kg (d) 4 kg
   (e) 1 kg (f) 5 kg

80

4.2a Grams

Objectives

- Estimate and weigh objects in grams.

Vocabulary

- Gram

Note

If students have trouble with reading scales, give them as much help as needed. Reading scaled number lines will be covered again in *Primary Mathematics* 2B in the context of bar graphs and does not have to be mastered yet. A gram does not have much weight. To estimate weight in grams, students should know how much some familiar objects weigh in grams. Prepare some objects in advance that are labeled with their approximate weights.

Introduce grams	
Tell students that the *gram* is the standard unit of measurement for objects lighter than a kilogram or between one kilogram and the next. Show students a unit cube or 2 paper clips and tell them that these weigh about 1 gram. You can pass some around so students can tell how light a gram is. Tell them to hold their pencils. Pencils are about 3 to 5 grams.	Paper clips Unit cube
Write **1 gram = 1 g** on the board and tell students that g is the abbreviation for grams.	1 gram = 1 g
Pass around some objects that are labeled with their weights to feel the heft.	
Interpret scales	Appendix pp. a 25-26
Draw a number line marked in 10's with 10 divisions between each major mark, or use copies of the ones in the appendix on p. a25. Mark some of the tick marks with an arrow pointing towards it. Have students count the divisions or spaces between the longer, labeled, tick marks. There are 10, so each stands for 1. Point to a tick mark and have students tell you what mark the arrow is pointing to stands for. The arrow at the right is pointing to 27. Now draw a number line marked in 100's, or use the one in the appendix. See if students can tell what each space from one tick mark to the next is now worth. Each is worth 10, so each tick mark is 10 more. Have students locate or label the correct mark for specific values, such as 270.	0 10 20 30 40 50 0 100 200 300 400 500
Give students copies of the circle scale on appendix p.a26. Tell them that it is like a number line wrapped in a circle. Tell them that each tick marks is 10 more along and that the longer ones between the hundreds are 50. Have them locate the marks that would represent specific values.	Appendix p. a26

Ask students to look at p. 68 in their textbooks to see the approximate weight in grams of some common items. Discuss the scales on p. 68 of the textbook and the divisions for the one on the left. Tell students that the one on the right is an electronic scale, which can weigh things more accurately.	**Text p. 68**
Estimate and weigh in grams	
Divide students into groups and have them estimate and weigh some items using a balance and weights of various grams, such as 10 g, 20 g, 50 g. How precise they can be in the weights depends on the balance. Finding the weight using various combinations of different weights, if you have them, requires more problem solving than just using a measuring scale. If you have enough weighing scales, let them also weigh a few objects using those. Make sure they can interpret the marks on whichever scales you have. They can fill in the table in problem 1 of Exercise 30 in the workbook, or you can provide them with a different chart using objects you are having them weigh.	Object | Estimate | Weight (blank chart with three empty rows)
Assessment	**Text p. 69**
	1. (a) 400 g (b) 600 g 2. (a) 350 g (b) 230 g 3. Answers will vary.
	WB Exercise 30, pp. 81-82, problems 2-3

Object	Estimate	Weight

Exercise 30

1. Answers will vary.

2. (a) 130 g
   (b) 90 g
   (c) 220 g
   (d) 40 g

81

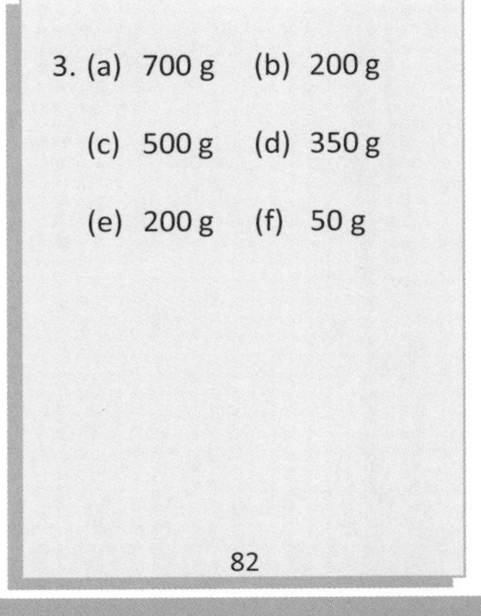
3. (a) 700 g (b) 200 g
   (c) 500 g (d) 350 g
   (e) 200 g (f) 50 g

82

4.3 4.4 Measuring Weight in Pounds and Ounces

Objectives

- Estimate and measure weight in pounds.
- Estimate and measure weight in ounces.
- Compare a pound to a kilogram.
- Compare an ounce to a gram.
- Review all topics.

Material

- Balance
- Pound and ounce weights
- Kilogram weights
- Scale marked in pounds and ounces
- Bag of 58 paper clips
- Charts (see lesson)
- Appendix p. a27

Notes

In Part 3 students will become familiar with the pound as a unit of measurement in the U.S. customary system. In Part 4 they will learn about ounces.

There are 16 ounces in a pound.

Students may be more familiar with pounds and ounces than kilograms and grams from everyday usage.

Students will later have word problems involving the weights of people. Weights in kilograms may be confusing, since a full-grown man weighs as much in kilograms as one of them might in pounds.

Since 1 kg = 2.205 pounds, a good approximation is 1 kg = 2 pounds. A person weighing 30 kg weighs about 60 pounds. A weight of 80 kg means an adult, since that is about 160 pounds. A small child could weigh 15 kg, or 30 pounds.

If you don't want to have students again use beans to make a weight, this time for 1 pound, grocery stores in the U.S. often sell beans in 1 pound bags.

If you don't have enough ounce weights for students to weigh with, you can tape together 11 pennies to make approximate ounce weights, or see how many counters or multilink cubes weigh about an ounce.

The abbreviation **lb** for pound comes from *libra*, the Latin word for "a balance scale" and "a unit of weight on the scale." In the U.S. # is sometimes used for pound. The abbreviation oz for ounce comes from *onza*, meaning "a unit of."

If you have weighing scales for use in the classroom, the scale on them may not be exactly like those in the text. Make sure students can interpret the scale on them if they are using scales to weigh objects.

Blank Page

4.4a Pounds and Ounces

Objectives

- Estimate and measure weight in pounds.
- Estimate and measure weight in ounces.
- Compare a pound to a kilogram.
- Compare an ounce to a gram.

Vocabulary

- Pounds
- Ounces

Note

The pages on pounds and ounces do not include new concepts, other than two more units of weight. You can cover both weights in one lesson, or spend an extra lesson on this material, if students are doing a lot of weighing.

To help with estimating weights, students should use a familiar object that weighs about a pound or an ounce or know the weight of some familiar objects in pounds or ounces. Five quarters weigh about an ounce.

Introduce pounds	
Tell students that in the United States we often use *pounds* as the customary unit of weight. For example, they probably know their own weight in pounds. They might know a sibling's weight at birth. They might have heard parents moaning about their weight. At the airport, there is a weight limit on baggage, usually 50 pounds. We usually buy 1-pound or 5-pound boxes or bags of sugar or flour in the store. The price for fruit is usually given as the cost for each pound. Pass around a pound weight so students can feel how heavy it is. See if they can tell you if it is heavier or lighter than a kilogram. Then let them compare it to a kilogram weight. You can use a balance to show that a kilogram is heavier than a pound. Tell them that 1 kilogram is slightly heavier than two pounds.	Pound weight Kilogram weight
Write **1 pound = 1 lb** on the board and tell students that **lb** is the abbreviation for pound. It is an odd abbreviation (as are the customary units of measurement as a whole), so you might want to discuss its origins to help students remember it.	1 pound = 1 lb
Introduce ounces	
Tell students that we weigh lighter objects in *ounces*. Pass some ounce weights around for students to feel. Show them 2 paper clips and ask them about how much they weigh (1 gram). Then show them a bag of about 58 paper clips and tell them it weighs about an ounce. So an ounce is quite a bit heavier than a gram, but still pretty light.	Ounce weight
Write **1 ounce = 1 oz** on the board and tell students that **oz** is the abbreviation for ounce. Tell them that there are 16 ounces in a pound. So scales that weigh up to 1 pound go up to 16 ounces, and those that weigh more than one pound may have 16 divisions between each pound. (If they have only two, then the mark between pounds is for 8 ounces, not 5.)	1 ounce = 1 oz 1 lb = 16 oz

Estimate and weigh in pounds or ounces

Text p. 70

In order for all groups to have pound weights, you may want to have them do the activity in Task 1 on p. 70 of the textbook. Or you can use pound bags of beans from the store.

Provide students with balances and pound and ounce weights and have them weigh various objects to the nearest pound or to the nearest ounce. You can prepare a chart as in previous lessons for them to record their results. Ask them to first estimate the weights.

You can also let the groups use measuring scales. Make sure they know how to read the scales and what each division represents.

Balances
Pound weights
Ounce weights

Object	Estimate	Weight

Assessment

Text pp. 71-73

2. heavier
3. (a) the one on the left; 2 lb
   (b) 16 lb
4. Answers will vary.
1. 6 oz   10 oz

Use the results from their weighing various objects to ask students some word problems where they add weights or find out how much more one item weighs than another.

Point to various objects and ask students if we would weigh them in pounds or ounces.

Practice

Appendix p. a27

1. (a) oz
   (b) lb
   (c) oz
   (d) oz
   (e) lb
2. 200 ounces
3. 8 oz: 2
   4 oz: 1
   1 oz: 3
4. 35 lb + 18 lb = 53 lb
   There were 53 lb at first.
5. 64 lb – 16 lb = 48 lb
   Jorge weighs 48 lb.

4.4b Practice and Review

Objectives

- Practice addition and subtraction within 1000.
- Solve word problems involving measurement.
- Review all topics.

Note

You may want to spend several days on this lesson, having students work on the practice one day and the review the next day, or you can save some of the problems and have students solve one or two during later lessons for more continuous review.

A durian (problem 6 of Practice 4A) is a strong-smelling large fruit common in Singapore.

Practice	Text p. 74, Practice 4A	Text p. 75, Review A
Students can do these problems individually and then share and discuss their solutions to the word problems. They can draw number bonds only if needed. Suggested number bonds are given below for problem 9 of Practice 4A. 9. (a) Apple and Pineapple 840 g Apple 90 g — Pineapple ? 9. (b) Pineapple 750 g Apple 90 g — ?	1. (a) 261 (b) 408 (c) 533 2. (a) 637 (b) 856 (c) 930 3. (a) 193 (b) 287 (c) 320 4. (a) 32 (b) 480 (c) 586 5. (a) 554 (b) 623 (c) 535 6. (a) The **durian** is heavier. (b) 900 g – 550 g = **350 g** The durian is 350 g heavier. 7. (a) 39 kg + 28 kg = **67 kg** His father weighs 67 kg. (b) 39 kg + 67 kg = **106 kg** Their total weight is 106 kg. 8. (a) 280 g – 60 g = **220 g** The pear weighs 220 g. (b) 280 g + 220 g = **500 g** The total weight is 500 g. 9. (a) 840 g – 90 g = **750 g** The pineapple weighs 750 g. (b) 750 g – 90 g = **660 g** The pineapple is 660 g heavier than the apple.	1. (a) 659 (b) 715 (c) 850 2. (a) 977 (b) 660 (c) 1000 3. (a) 402 (b) 782 (c) 810 4. (a) 500 (b) 350 (c) 32 5. (a) 184 (b) 625 (c) 398 6. 20 m – 7 m = **13 m** 13 m of the rope was left. 7. 96 cm + 85 cm = **181 cm** She used 181 cm. 8. 340 g – 95 g = **245 g** The papaya weighs 245 g. 9. (a) 34 kg – 8 kg = **26 kg** Her brother weighs 26 kg. (b) 34 kg + 26 kg = **60 kg** Their total weight is 60 kg.

Reinforcement	
Write each pair of measurements shown below on the board, or say them aloud, and ask students which is greater.	
⇒ 1 lb, 1 kg ⇒ 3 cm, 3 m ⇒ 4 oz, 4 g ⇒ 8 in., 8 cm ⇒ 30 m, 30 yd ⇒ 16 oz, 1 lb ⇒ 16 in., 1 ft ⇒ 12 cm, 1 ft ⇒ 12 in., 1 ft	1 lb < 1 kg 3 cm < 3 m 4 oz > 4 g 8 in. > 8 cm 30 m > 30 yd 16 oz = 1 lb 16 in. > 1 ft 12 cm < 1 ft 12 in. = 1 ft
Review	WB Review 3, pp. 83-86 WB Review 4, pp. 87-90

Review 3

1. (a) 192 (b) 209
   (c) 370 (d) 405
   (e) 66 (f) 605
   (g) 398 (h) 909
2. (a) 90
   (b) 7
   (c) 700
   (d) 200
3. 62 + 38 = 100
   38 + 62 = 100
   100 – 38 = 62
   100 – 62 = 38

83

4. (a) 389 (b) 500
   (c) 416 (d) 402
   (e) 1000 (f) 55
5. (a) 528
   (b) 369
   (c) 951
   (d) 369
   (e) 634

84

6. 285 + 167 = 452
   452
7. 635 + 165 = 800
   800
8. 600 – 485 = 115
   115

85

9. 148 + 137 + 359 = 644
   644
10. $305 - $59 = $246
    $**246**
11. 580 + 85 = 665
    665

86

Review 4

1. pear shape
2.

greatest number	smallest number
420	204
431	**134**
754	**457**
432	**234**
954	**459**
330	**303**

87

3. (a) 689
   (b) 40
   (c) 80
   (d) 200
4. 12 cm
   10 cm
5. (a) 130 g
   (b) 210 g – 130 g = 80 g
   80 g

88

6. 312 - 295 = 17
   17
7. 292 + 149 + 68 = 509
   509
8. $502 - $348 = $154
   $**154**

89

9. $650 - $527 = $123
   $**123**
10. 860 g - 280 g = 580 g
    580 g
11. 34 cm + 28 cm + 16 cm
    = 78 cm
    78 cm

90

5 Multiplication and Division

Objectives

- Write multiplication equations for equal groups.
- Use repeated addition to evaluate a multiplication equation.
- Use rectangular arrays to illustrate multiplication.
- Write two related multiplication equations for an array.
- Write division equations.
- Divide by sharing equally into a given number of groups.
- Divide by grouping by a specified number to find the number of groups.
- Use rectangular arrays to relate division to multiplication.
- Write two multiplication and two division equations for a given set of equal groups.
- Solve word problems which involve multiplication and division.

Suggested number of weeks: 2

		TB: Textbook WB: Workbook	Objectives	Material	Appendix
5.1	**Multiplication**				
5.1a	Equal Groups	TB: pp. 76-77 WB: pp 91-92	♦ Write multiplication equations for equal groups. ♦ Use repeated addition to evaluate a multiplication equation.	♦ Counters ♦ Multilink cubes	
5.1b	Multiply	TB: p. 78 WB: pp. 93-94	♦ Multiply a number by another.	♦ Counters ♦ Multilink cubes ♦ Number cubes	
5.1c	Arrays	TB: p. 78 WB: pp. 97-98	♦ Use rectangular arrays to illustrate multiplication. ♦ Write two related multiplication equations.	♦ Counters ♦ Multilink cubes ♦ Graph paper ♦ Number cubes	
5.1d	Word Problems	TB: p. 79 WB: pp. 95-96	♦ Solve simple word problems which involve multiplication.	♦ Counters ♦ Multilink cubes	

		TB: Textbook WB: Workbook	Objectives	Material	Appendix
5.2	**Division**				
5.2a	Sharing	TB: pp. 81-82 WB: pp. 99-102	♦ Divide by sharing equally into a given number of groups. ♦ Write division equations. ♦ Solve simple word problems which involve sharing.	♦ Counters ♦ Paper cups or plates	a28
5.2b	Grouping	TB: pp. 80, 83-84 WB: pp. 103-106	♦ Divide by grouping by a specified number to find the number of groups. ♦ Solve simple word problems which involve grouping.	♦ Counters ♦ Multilink cubes ♦ Craft sticks	a29
5.2c	Arrays	TB: p. 85 WB: pp. 107-109	♦ Use rectangular arrays to relate division to multiplication. ♦ Write two division equations and two multiplication equations for a given set of equal groups.	♦ Counters ♦ Multilink cubes	
5.2d	Practice and Review	TB: pp. 86-87 WB: pp. 110-113	♦ Solve word problems which involve multiplication and division. ♦ Review all topics.	♦ Counters	a30

5.1 Multiplication

Objectives

- Write multiplication equations for equal groups.
- Use repeated addition to evaluate a multiplication equation.
- Use rectangular arrays to illustrate multiplication.
- Write two related multiplication equations for an array.
- Solve simple word problems which involve multiplication.

Material

- Counters
- Multilink cubes
- Number cubes 1-6, 1 per group
- Graph paper

Prerequisites

In order for students to find answers to multiplication problems using repeated addition, they need to be comfortable with mental addition of a 1-digit number to a 2-digit number. This is covered in *Primary Mathematics* 1B and reviewed in the lessons for Unit 2, Part 1. If they are not comfortable with mental addition, they can count up. In *Primary Mathematics* 1, students learned to count by 2's, 5's, and 10's, which they can use to find the answer to multiplication problems where there are 2, 5, or 10 objects in each group.

Notes

Students were introduced to the concept of multiplication and used repeated addition to find the value of multiplication expressions in *Primary Mathematics* 1B. This is reviewed here.

Multiplication is associated with the part-whole concept. Given the number of equal parts and the number in each part, we can multiply to find the whole (the total).

In *Primary Mathematics* 1B students wrote 3 groups of 4 as 3 x 4. Here, they will learn that they can write it as either 3 x 4 or 4 x 3 (commutative property of multiplication). 3 groups of 4 and 4 groups of 3 give the same answer, and the order of the factors is not important in the written expression. 3 x 4 could just as well be read as "3 in 4 groups" or "3 multiplied by 4." By the time students learn algebra, there will be no arbitrarily imposed order to the factors according to which one is the number of parts and which one is the number in each part.

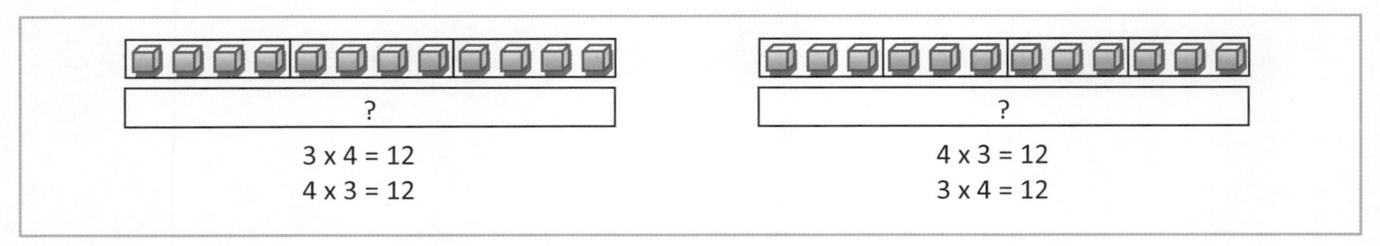

Note that in Task 1 on p. 77 of the text, there are 5 birds in 4 nests, 5 x 4. The number of groups is the second factor. In Task 2, there are 5 groups of 6 hats, 5 x 6. The number of groups is the first factor.

Rectangular arrays will be used to further illustrate the commutative nature of multiplication. With an array, students can visually see that the answer is the same whether the rows or the columns are considered the group.

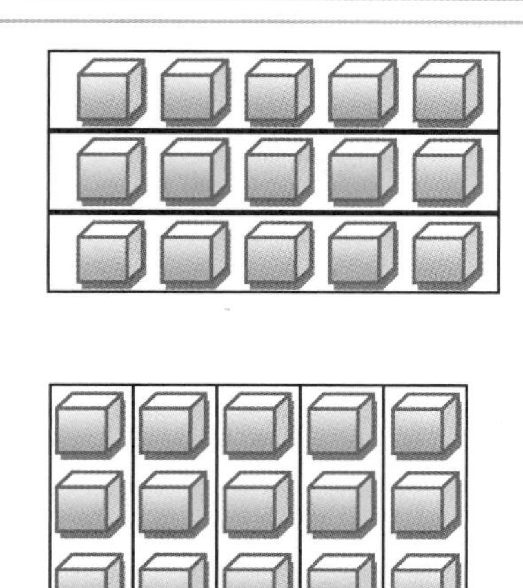

3 rows of 5
5 + 5 + 5
3 x 5 = 15
A row of 5 multiplied by 3.
5 x 3 = 15

5 columns of 3
3 + 3 + 3 + 3 + 3
5 x 3 = 15
A column of 3 multiplied by 5.
3 x 5 = 15

Students will be asked to explore ways they can form objects into arrays. An array can have only one row or one column (e.g., 1 x 12 = 12 or 12 x 1 = 12). Students will explore the identity nature of 1 for multiplication in *Primary Mathematics* 3A, along with multiplication by 0.

When we are simply given an isolated expression, such as 3 x 4, without a word problem to go with it, it is not possible or necessary to interpret which factor is the number in each group and which is the number of groups in order to evaluate the expression. Students can solve this either by adding 3's 4 times, or adding 4's 3 times. If they need to illustrate a problem or use manipulatives to find the answer, accept either factor as the group.

Later, students will learn to interpret the word *of* to mean multiplication, as in $\frac{3}{4}$ *of* 24 or 15% *of* 100. You can prepare them for this by emphasizing the word *of*, or even sometimes leave out groups, such as 3 *of* 5's instead of 3 groups of 5.

The emphasis in this unit is on understanding the concept of multiplication. Students are not required to memorize multiplication facts yet. They can solve the problems in this unit with manipulatives, pictures or repeated addition. Do not require them to constantly verify their answer with pictures if they can find the answer with repeated addition, or even if they have already learned (memorized) a multiplication fact.

Students should know how to count by 2's, 5's, and 10's by now. They will learn to count by 3's in the next unit. Do not spend time teaching students to count by 4's, 6's, 7's, 8's or 9's by memory at this time. It is not necessary for learning multiplication facts — students will be given other strategies in *Primary Mathematics* 2B and 3A.

5.1a Equal Groups

Objectives

- Write multiplication equations for equal groups.
- Use repeated addition to evaluate a multiplication equation.

Vocabulary

- Equal groups
- Of

Note

Students have previously written 3 x 5 for *3 groups of 5*. They will now learn that it can also be written 5 x 3. You can make it into an English sentence, such as, *5 in 3 groups*, or *5 multiplied by 3*, as in the next lesson, but ultimately, the equation can simply be read as *three times five* regardless of the situation it is describing.

Students will probably find answers by simply counting on from the amount in the first group if the problem supplies a picture. Encourage them to add repeatedly using mental math.

Equal groups	
Show students some objects arranged in 3 equal groups of 5. You can use counters or multilink cubes or draw them on the board. Ask students how many groups there are and how many counters there are in each group. Point out that these are *equal groups*. Draw a number bond with 3 parts. Ask students how we can find the total number. We can add: 5 + 5 + 5 = 15. 5 and 5 is 10, then 10 and 5 is 15. Ask students if they remember the symbol we can use when we add equal groups. It is the multiplication symbol, "x." Tell them we write 3 x 5 for *3 groups of 5* or *3 of fives* or *3 fives*, or we can write 5 x 3 for *5 in 3 groups*. As with addition, it does not matter what order we write the 5 and the 3.	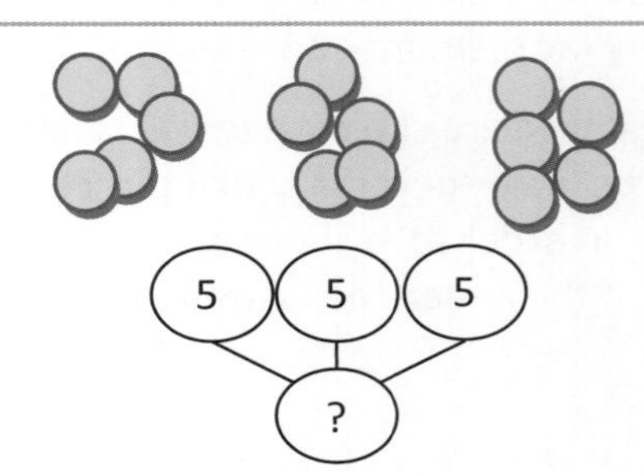 5 + 5 + 5 = 15 3 x 5 = 15 5 x 3 = 15
Repeat, this time with 5 groups of 3. Students should see that the answer is the same as with 3 groups of 5. The multiplication equations used are also the same, and it does not matter which number is written first.	3 + 3 + 3 + 3 + 3 = 15 5 x 3 = 15 3 x 5 = 15
Discussion	**Text pp. 76-77**
Discuss textbook p. 76 and the top of p. 77. Note that the girl thinks "3 x 4 = 12" and the boy thinks "4 x 3 = 12" for 3 groups of 4. Both are correct. Tasks 1 and 2: Encourage students to add mentally, rather than count on one by one.	There are **12** apples altogether. 1. 20 2. 30

Addition and multiplication	
Provide students with counters or multilink cubes. Students can work in groups. Each group should have 50 counters or cubes. Write the two expressions 4 + 3 and 4 x 3 on the board and ask students to illustrate each of them with their counters. See if a student can explain the difference. 4 + 3 means there are two parts, 4 and 3, that are added together. 4 x 3 means there are equal parts, either 4 equal parts with 3 in each part, or 3 equal parts with 4 in each part.	4 + 3 4 x 3
Write the two expressions 4 + 4 and 4 x 4 and ask students to illustrate them with their cubes. We can still have equal parts with addition. For 4 + 4 we have two parts. In multiplication, we must have equal parts. For 4 x 4, we have 4 equal parts. Ask students if they can write 4 + 4 using a multiplication sign. Write 4 + 4 = 4 x 2 (or 4 + 4 = 2 x 4) on the board. Remind students that the equal sign means that the expression on both sides have the same value.	4 + 4 4 x 4 4 + 4 = 4 x 2
Assessment	
Use counters or draw a picture on the board of equal groups and get students to write a multiplication expression. Write some multiplication expressions on the board and have students find the answers. They can add mentally, use the cubes, or draw pictures to find the answers. Do not require students to always draw pictures if they don't need to. Limit the problems to ones where the product is no more than 50.	
Practice	WB Exercise 31, pp. 91-92

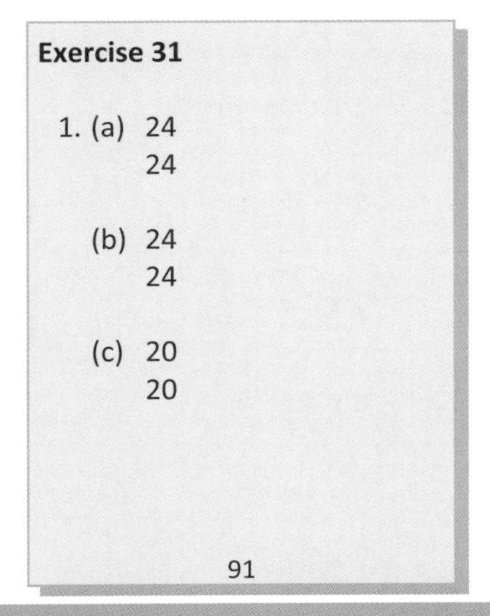

Exercise 31

1. (a) 24
   24

   (b) 24
   24

   (c) 20
   20

91

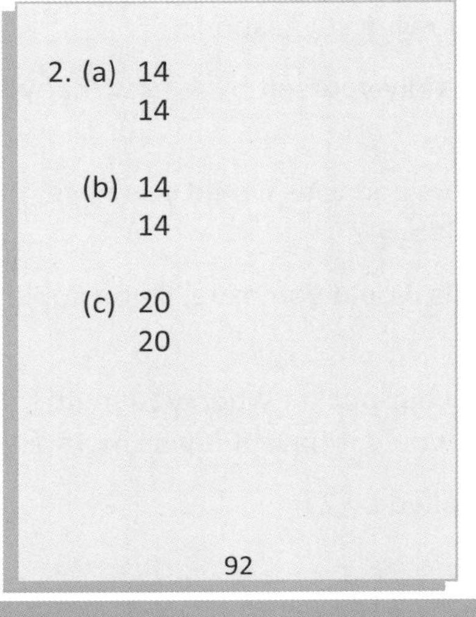

2. (a) 14
   14

   (b) 14
   14

   (c) 20
   20

92

5.1b Multiply

Objectives

- Multiply a number by another.

Vocabulary

- Multiply

Note

In the previous lesson students looked at multiplication as adding equal groups. 4 groups of 5 is most often written as 4 x 5. Here, they will consider multiplication as a more dynamic process; a group or set multiplied by a given number. A group of 5 multiplied 4 times is most often written as 5 x 4. In the next lesson, they will see that both ideas can be represented by an array, which will help them further understand the commutative nature of multiplication.

Multiply	
Discuss the concept of *multiply* with students. To multiply means to increase the number of something, generally by copying it. For example, animals and plants multiply to propagate themselves; that is, make more of the same species. Maybe they have wished they could be more than one place at a time, and could multiply themselves. Perhaps they have heard a parent complain of not having enough hands; if only they could multiply their hands. Ask students how many hands they would have if they could multiply themselves into four people. Both hands are multiplied by four. Since there are 2 hands, 2 is multiplied by 4.	
Tell students that in mathematics, quantities are multiplied into equal groups (doubled, tripled, quadrupled, etc.). Set out 2 counters and write the expression 2 x 4. Read it as 2 multiplied by 4. Add 3 more pairs of counters. Now 2 is multiplied by 4; that is, 2 is multiplied so that there are now four 2's. Ask students how many there are now. When we multiply 2 by 4, we have 8.	2 multiplied by 4 2 x 4 = 8
Provide students with counters or multilink cubes. Give them other examples, having them work out the answers and write the equations. If using the example of multiplying yourself, you could use some of the following suggestions:	Counters or multilink cubes
⇒ If you could multiply yourself by 4, how many fingers would you have?	10 x 4 = 40
⇒ How many fingers and toes would you have, if you could multiply yourself by 4?	20 x 4 = 80
⇒ How many heads would you have, if you could multiply yourself by 4?	1 x 4 = 4
⇒ A spider was in your pocket when you multiplied yourself — how many legs would it have if it was multiplied 4 times?	8 x 4 = 32
⇒ What is 6 multiplied by 2?	6 x 2 = 12
⇒ Multiply 3 by 8.	3 x 8 = 24

Assessment	Text p. 78
	3. (a) 21 (b) 36
Practice	WB Exercise 32, pp. 93-94

Group Game

Material: Number cube 1-6, one per group.

Single player: Roll the number cube twice. After the first roll, draw the corresponding number of circles. After the second roll, draw X's in each circle. Write the multiplication sentence for the total number of X's. For example, a 4 and a 5 are rolled. 4 x 5 = 20

Multi-player: Each player rolls the number cube and draws the corresponding number of circles. Then each player rolls the cube a second time, and draws the corresponding number of X's in each circle. Then they each write a multiplication equation and find the total number of X's, For example, a 4 and a 5 are rolled. 4 x 5 = 20. The player with the most X's wins the round.

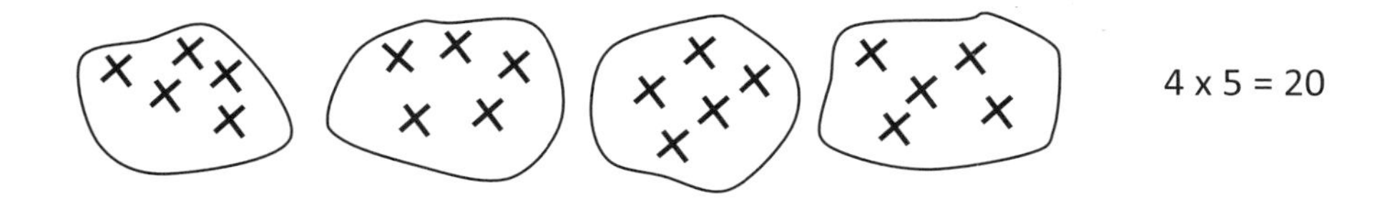

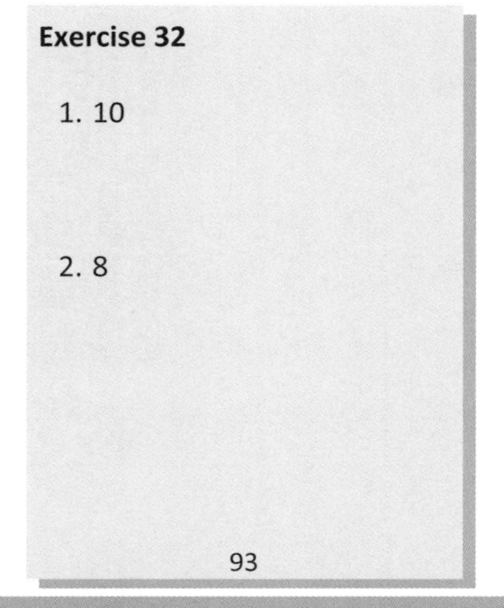
Exercise 32

1. 10

2. 8

93

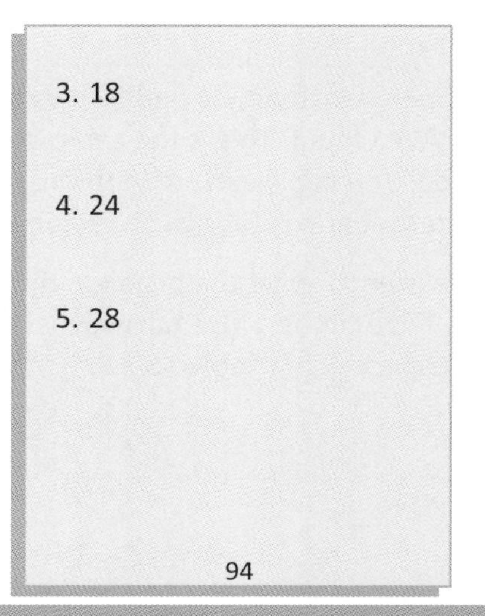
3. 18

4. 24

5. 28

94

5.1c Arrays

Objectives

- Use rectangular arrays to illustrate multiplication.
- Write two related multiplication equations.

Vocabulary

- Array
- Row
- Column

Note

Workbook Exercise 34 is assigned before Exercise 33. If you want to keep the order, assign both after the next lesson on word problems.

Arrays	
Use multilink cubes or counters or draw a picture to show a 5 by 3 array. Tell students that an *array* is an arrangement of objects in *rows* and *columns*. Have them identify the rows and the columns. Ask students how many rows there are and how many blocks there are in each row. We can think of the rows as equal groups, so we have 3 rows of 5. We can also think of this as a row of 5, multiplied by 3.	3 rows of 5 5 + 5 + 5 3 x 5 = 15 A row of 5 multiplied by 3. 5 x 3 = 15
Then ask students how many columns there are and how many blocks are in each column. We can think of the columns as equal groups, so we have 5 columns of 3. We can also think of this as a column of 3, multiplied by 5. Write the multiplication equations. Tell students that however we think of the arrangement, or write the equations, there are always 15 blocks. We can write either 5 x 3 or 3 x 5 for this array.	5 columns of 3 3 + 3 + 3 + 3 + 3 5 x 3 = 15 A column of 3 multiplied by 5. 3 x 5 = 15
Ask what would happen if instead we had 5 rows of 3. Would we still have 15 blocks? We would. This is the same array as before, just turned on its side. You can illustrate with the multilink cubes by sticking them together and rotating the rectangular shape. Ask students if it is easier to solve the problem by adding 5 three times, or by adding 3 five times. Since both give the same answer, we can choose the easiest way to solve 5 x 3.	3 x 5 = 5 x 3 = 15

Create arrays	
Provide students with counters or multilink cubes. Ask them to create an array, such as one with 5 rows of 6, and write two multiplication equations. Repeat with a few other arrays.	5 x 6 = 30 6 x 5 = 30
Assessment	**Text p. 78**
	4. (a) 4 x 2 = **8** 2 x 4 = **8** (b) 5 x 3 = **15** 3 x 5 = **15**
Practice	WB Exercise 34, pp. 97-98

Group Game

Material: Number cube 1-6, one per group, graph paper or multilink cubes for each player.

Procedure: Players take turns rolling the number cube twice. The first roll is the number of squares in a row, and the second number the number of rows. Students can either color squares on the graph paper to make the rectangle or make the array with cubes. They then write two multiplication equations. For example, a 6 and a 3 are rolled. The number of squares in the rectangle is the score. Each player rolls a specified number of times (such as 6 times to make 3 separate arrays) and the scores are added. The player with the highest score wins.

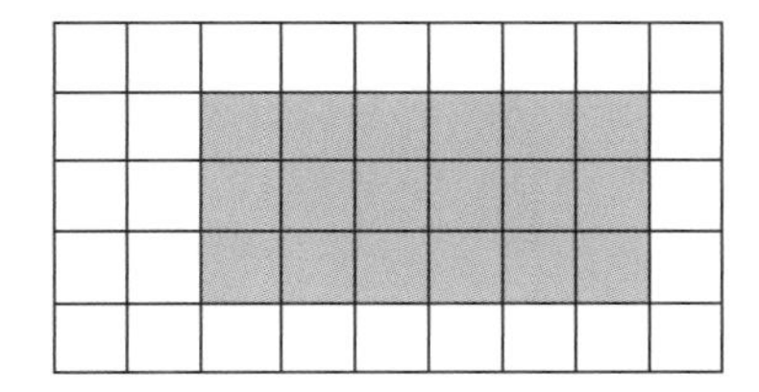

6 x 3 = 18
3 x 6 = 18

Exercise 34

1. (a) 15 15

(b) 14 14

(c) 18 18

97

2. (a) 30 30

(b) 28 28

(c) 48 48

(d) 30 30

98

5.1d Word Problems

Objectives

- Solve simple word problems which involve multiplication.

Note

In the next unit, students will not always have pictures to go with the problems, and should draw pictures or number bonds if needed and move away from using manipulatives. In *Primary Mathematics* 3, students will draw bar models for word problems. For problems that involve multiplication, they will draw bars with equal parts. In preparation for this, you can draw some multi-part number bonds for some of these problems.

Students can write the factors in any order. Only one equation is given for the sample problems here.

Word problems	
Provide students with counters or multilink cubes. Write some multiplication word problems on the board, such as the ones given below. Include a few addition problems. Students can use the counters to find the answer, or draw a simple picture, as shown for the first problem. Ask them to write a multiplication equation for each, if the problem involves equal groups, or an addition equation if the problem involves adding two parts. As you discuss the problems, point out that we need to find a whole, or total. Then ask if we are given unequal parts or equal parts. If there are equal parts, ask how many are in each part.	Counters or multilink cubes
⇒ If a stool has 3 legs, how many legs are there on 6 stools?	3 3 3 3 3 3 ? 3 x 6 = 18 or 6 x 3 = 18
⇒ Harriet's birds all have two wings. How many wings do 8 of her birds have?	2 x 8 = 16 8 birds have 16 wings.
⇒ Macaws have 4 toes. How many toes do 6 macaws have?	4 x 6 = 24 6 macaws have 24 toes.
⇒ How many toes do 4 children have?	4 x 10 = 40 4 children have 40 toes.
⇒ How many fingers are there on 2 hands?	5 x 2 = 10; or 5 + 5 = 10 There are 10 fingers on 2 hands.
⇒ How many fingers are there on 5 hands?	5 x 5 = 25 There are 25 fingers on 5 hands.
⇒ How many legs do 3 insects have?	3 x 6 = 18 3 insects have 18 legs.

⇒ How many legs do 5 spiders have?	5 x 8 = 40 5 spiders have 40 legs.
⇒ There are 3 insects and 5 spiders. How many legs are there altogether? How many heads are there in all?	18 + 40 = 58 3 insects and 5 spiders have 58 legs and 8 heads.
⇒ There are 4 insects and 3 spiders now. How many legs are there in all?	4 x 6 = 24; 3 x 8 = 24 24 + 24 = 48 4 insects and 3 spiders have 48 legs.
⇒ I saw 9 people dangling their feet in a stream. A dog hid their shoes. How many shoes did she hide?	9 x 2 = 18 The dog hid 18 shoes.
Assessment	**Text p. 79, Practice 5A**
	1. 6 x 4 = 24 or 4 x 6 = **24** There are 24 butterflies. 2. 2 x 5 = 10 or 5 x 2 = **10** There are 10 buttons on 5 dresses. 3. 6 x 3 = 18 or 3 x 6 = **18** There are 18 chairs in 3 rows. 4. 3 x 5 = 15 or 5 x 3 = **15** There are 15 cakes altogether.
Write a multiplication expression on the board and have students find the answer and make up a word problem to go with it.	
Practice	WB Exercise 33, pp. 95-96

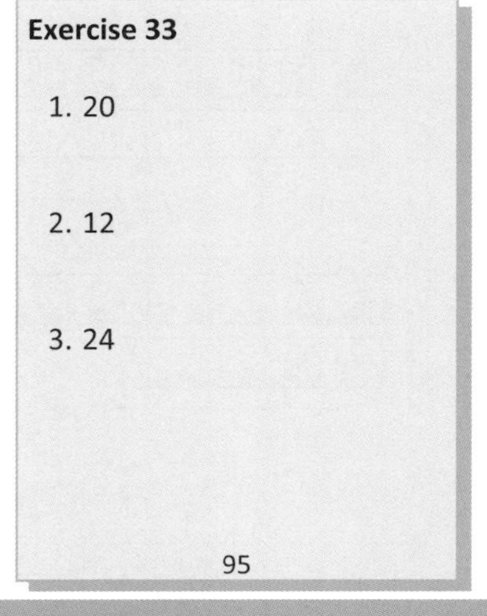

Exercise 33

1. 20

2. 12

3. 24

95

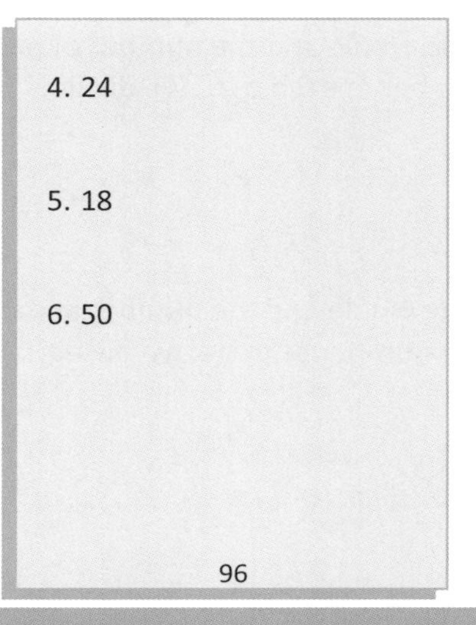

4. 24

5. 18

6. 50

96

5.2 Division

Objectives

- Write division equations.
- Divide by sharing equally into a given number of groups.
- Divide by grouping by a specified number to find the number of groups.
- Use rectangular arrays to relate division to multiplication.
- Write two multiplication and two division equations for a given set of equal groups.
- Solve word problems which involve multiplication and division.

Material

- Counters
- Multilink cubes
- Paper cups or plates
- Craft sticks
- Appendix pp. a28-a30

Notes

Students were introduced to the concept of division in *Primary Mathematics* 1B. This part reviews the concept and introduces the division symbol. The emphasis is on understanding the meaning of division rather than on memorization of division facts or on finding division facts from multiplication facts. Let students use pictures or manipulatives to find the answers to division problems in this part.

Page 80 in the textbook illustrates two kinds of division situations:

Sharing:
Start with a set of objects (12 balloons).
Make a given number of equal groups (3 groups).
Find the number of objects in each group (4 balloons).

Grouping:
Start with a set of objects (12 balloons).
Make equal groups of a given size (4 balloons).
Count the number of groups made (3 groups).

Previously students learned that addition and subtraction are associated with the part-whole concept. If we are given two parts, we can add to find the whole. If we are given the whole and a part, we can subtract to find the other part. Multiplication and division are also associated with the part-whole concept. Instead of two or more different parts making a whole, a specified number of equal parts make the whole.

If we are given the number of equal parts and the number in each part and need to find the whole (total), we multiply.

If we are given the whole and the number of parts and need to find the number in each part, we divide.

If we are given the whole and the number in each part and need to find the number of parts, we divide.

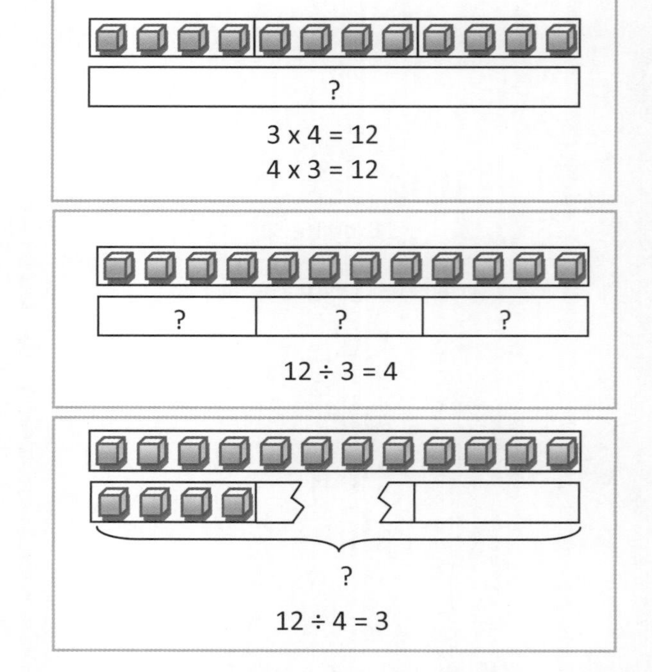

In *Primary Mathematics* 3, students will be introduced to the term "unit" for the equal parts. At this level, they should understand that both multiplication and division are associated with equal parts.

Division by grouping can be thought of as repeated subtraction. Repeated subtraction, however, is not a very helpful way of doing division in the long run since it is necessary to keep track of how many times the number is subtracted. A concept of "repeated subtraction" is therefore not emphasized in this curriculum. The essential concept here is that when we know how many go into each group, and want to find the number of groups, we use division.

Arrays will be used to provide a visual way to see the relationship between the two division situations and between multiplication and division.

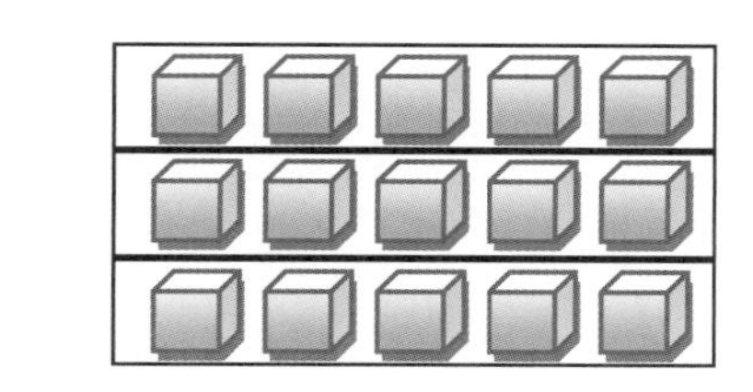

3 x 5 = 15
5 x 3 = 15
15 ÷ 3 = 5
15 ÷ 5 = 3

5.2a Sharing

Objectives
- Divide by sharing equally into a given number of groups.
- Write division equations.
- Solve simple word problems which involve sharing.

Vocabulary
- Share equally
- Divide
- Division
- Left over

Note

In this lesson, students will look at division as sharing. In the next lesson they will look at division as grouping. In the lesson after that, they will relate division to multiplication. Students may comment on the relationship between division and multiplication sooner than the third lesson.

Since not all numbers do share equally into a desired number of groups, these lessons will briefly include the concept of remainders.

Share equally	
Provide students with 15 counters and either paper cups or small paper plates. Ask them to put an equal number of counters onto each plate. One way to do this is to first put one on each plate, then a second on each plate, and so forth. Ask them how many are on each plate. Tell them we had a total amount, 15, and *shared* them *equally*, or *divided* them into 3 equal groups. This process is called *division*. Write the equation **15 ÷ 3 = 5**. Tell students that we use the division symbol, ÷, to show that we are dividing the 15. The number before the division symbol is the whole, or total amount. The number after the division symbol is the number of groups we are dividing into. The answer is how many go into each group. The equal sign means that the value on both sides is the same. Ask students why we cannot write the two numbers, 15 and 3, in any order. As with subtraction, we must always write the whole, or total amount first. That is the amount we are dividing up. We can't write 3 first, because we are not dividing up 3 things.	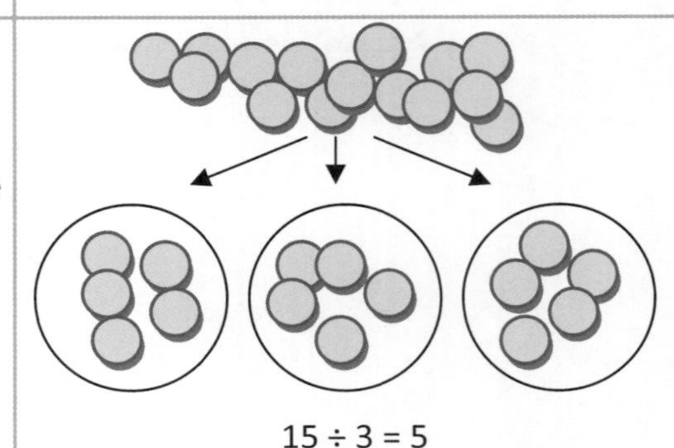 15 ÷ 3 = 5
Have students put the 15 counters together again and then divide them into 5 groups and write a division equation. Students may notice that the answer and the second number are switched around compared to the previous division equation. If they do, you can draw number bonds to illustrate both situations. See if they can see that division is the opposite of multiplication. In multiplication, we had equal groups and wanted to find the total. In division, we have a total and want to make equal groups. 3 groups of 5 are the same as 5 groups of 3. The number after the division symbol is the number of groups, and the number after the equal sign is the number in each group.	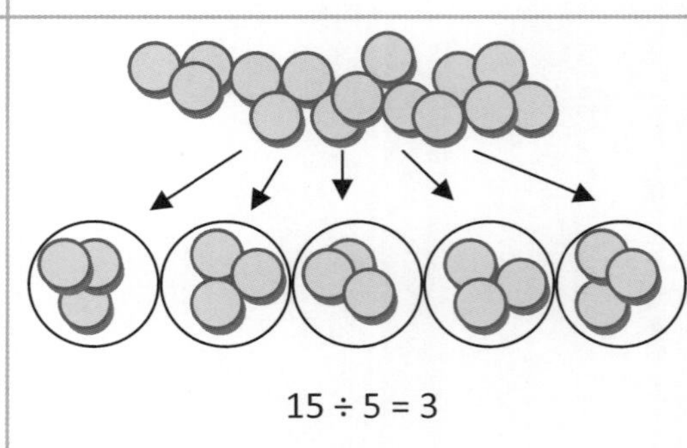 15 ÷ 5 = 3 15: 5, 5, 5 15: 3, 3, 3, 3, 3

Have students put the 15 counters together again and then divide them into 2 equal groups. They will find that there is one left over. Tell them that we cannot divide 15 into 2 equal groups. We can say that 15 divided by 2 is 7 with 1 left over.	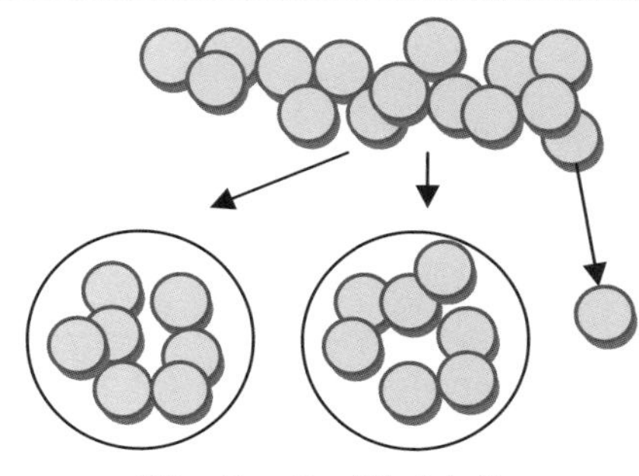 $15 \div 2 = 7$ with 1 left over
Word problems	Appendix p. a28
Use the problems on appendix p. a28 or similar problems that involve sharing. Students can use counters or draw simple pictures.	
1. Harriet bought some seed cakes for her birds. They ate 25 seed cakes in 5 days. How many seed cakes did they eat in one day?	$25 \div 5 = 5$ They ate 5 seed cakes in one day.
2. Harriet had 3 friends over to see her birds. She made 24 cookies. She and her friends shared the cookies equally. How many cookies did Harriet get?	$24 \div 4 = 6$ Harriet got 6 cookies.
Assessment	**Text pp. 81-82**
	2. 5; 5 3. 6; 6
Write some division expressions on the board and have students find the answers using counters. Write a division expression on the board and have students make up a word problem to go with it and find the answer.	
Practice	WB Exercise 35, pp. 99-100 WB Exercise 36, pp. 101-102

Exercise 35

1. (a) 6
   (b) 8
   (c) 4

99

2. (a) 5
   (b) 5
3. (a) 6
   (b) 6
4. 4

100

Exercise 36

1. (a) 6
   (b) 7
   (c) 3

101

2. $32 \div 4 = 8$
   8
3. $30 \div 6 = 5$
   5

102

5.2b Grouping

Objectives

- Divide by grouping by a specified number to find the number of groups.
- Solve simple word problems which involve grouping.

Note

Students may notice that the equation for 10, grouped by 2, gives 5 groups (10 ÷ 2 = 5). This is the same as the equation for 10, put into 2 groups, gives 5 in each group (10 ÷ 2 = 5). You can say that the numbers are the same, and the answer is the same, whether one means the number of groups and the other the number in each group. This will be discussed more in the next lesson.

Make equal groups	
Give students multilink cubes and ask them to count out 10 cubes. Then ask them to make a group of 2 from them, then another group of 2, and so on until they have grouped all the cubes. Ask them how many groups they have. (5) Tell students that when we group objects and then find the number of groups we are also dividing; we are making equal groups. Write the division equation and discuss the meaning of each term and symbol. ⇒ **10** is the total we started with. ⇒ ÷ means we are dividing. ⇒ **2** is the number that goes in each group. ⇒ = means that the expressions on both sides have the same value. ⇒ **5** is the number of equal groups.	10 ÷ 2 = 5
Have students count out another set of 10 cubes and group them by 5 and write the division equation. Discuss the meaning of each number and symbol again.	10 ÷ 5 = 2
Ask students to use another 10 cubes and group them by 3. This time, there will be one left over. We cannot make equal groups of 3 from 10. There will be 3 groups with one left over.	10 ÷ 3 = 3 with 1 left over

Discussion	**Text p. 80**
Ask students which child is sharing out the balloons (the boy) and which is making groups (the girl). Students can act out the two situations with counters. Ask them to write division equations. Discuss the meaning of each of the numbers in the two equations. In the equation for the boy, 12 balloons are shared into 3 equal groups. There are 4 balloons in each group. For the girl, 12 balloons are grouped by 4. There are 3 groups.	Boy: 12 ÷ 3 = 4 Girl: 12 ÷ 4 = 3
Word problems	Appendix p. a29
Use the word problems on appendix p. a29 or similar problems. Have students write a division expression and use counters or multilink cubes to find the answer.	
1. One night, Harriet wanted find out how many birds were on a perch without shining the flashlight into their faces. She counted 30 feet. How many birds were on the perch?	30 ÷ 2 = 15 There were 15 birds.
2. Harriet wants to take 24 birds to a bird show. She wants to put 3 birds in each cage. How many cages does she need?	24 ÷ 3 = 8 She needs 8 cages.
3. If she wants to take 26 birds instead, how many cages would she need? For 25 birds, how many cages would she need?	26 ÷ 3 = 8 with 2 left over. She would need 9 cages for 36 birds. She would also need 9 cages for 35 birds.
Assessment	**Text pp. 83-84**
	5. 5; 5 6. 6; 6
Practice	WB Exercise 37, pp. 103-104 WB Exercise 38, pp. 105-106

Activity

Material: 36 craft sticks per group.

Procedure: Ask students to see how many separate squares they can make from the sticks. Then ask them how many separate triangles they can make. Then ask them how many shapes with 5 sides they can make from the 3 sticks. Will there be any sticks left over? Have them see what other shapes they can make using 36 sticks without having any sticks left over.

Exercise 37

1. (a) 3
   (b) 4
   (c) 6

103

2. (a) 3
   (b) 3

3. (a) 7
   (b) 7

4. 3

104

Exercise 38

1. (a) 2
   (b) 4
   (c) 3

105

2. 18 ÷ 2 = 9
   9

3. 15 ÷ 3 = 5
   5

106

5.2c Arrays

Objectives

- Use rectangular arrays to relate division to multiplication.
- Write two division equations and two multiplication equations for a given set of equal groups.

Note

In this lesson, students will relate division to multiplication through the visual model of an array as well as a number bond with multiple equal parts. In this lesson counters are used, but you can use multilink cubes instead, or draw an array as a rectangle divided into rows and columns. Students should be able to visualize arrays as rows and columns with different objects, not just with cubes.

Multiplication and division	
Provide students with 12 multilink cubes or counters. Ask them to group them by 4 and write the division equation (12 ÷ 4 = 3). Point out that the answer is the number of groups. Draw a number bond to represent the situation. Then ask them what division equation they would write if they had been asked to share equally into 3 groups (12 ÷ 3 = 4). Point out that the answer now is the number in each group. For both division equations the first number is the whole. Whenever we divide we start with at total amount, or a whole. Ask students to write two multiplication equations for 3 groups of 4, or 4 multiplied by 3. Point out that with multiplication, the answer is the whole.	12 ÷ 4 = 3 3 x 4 = 12 12 ÷ 3 = 4 4 x 3 = 12
Repeat, but this time ask students to group the 12 objects by 3. The resulting equations are the same.	12 ÷ 3 = 4 4 x 3 = 12 12 ÷ 4 = 3 3 x 4 = 12
Have students arrange their counters into an array, with rows and columns. It does not matter if there are 3 rows and 4 columns or 4 rows and 3 columns. Some students can do it one way, and some the other. Point out that either the rows or the columns could be the equal groups. Tell them we can write two related division equations. We start with the whole and divide. The whole is always written first. We can also write two related multiplication equations. The answer is the whole that we divided up in the division problems.	
Get students to arrange the 12 objects in different arrays and write four multiplication and division equations for each. Point out that there are some arrays they can't make with 12, such as equal rows of 5. Point out that we can have a single row or column in an array. 12 objects grouped by 12 gives 1 group.	12 ÷ 1 = 12 12 ÷ 2 = 6 12 ÷ 12 = 1 12 ÷ 6 = 2 1 x 12 = 12 6 x 2 = 12 12 x 1 = 12 2 x 6 = 12

Repeat with 11 objects. Students will find that they can only make one row or column, any other arrangement results in unequal groups.	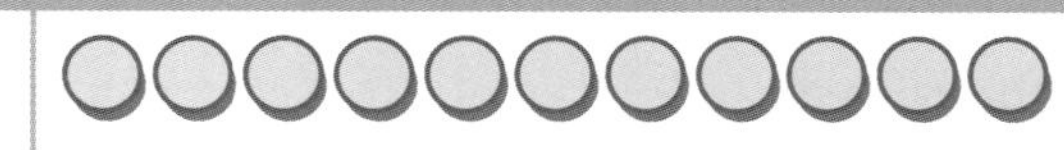 11 ÷ 1 = 11 11 ÷ 11 = 1 1 x 11 = 11 11 x 1 = 11
Four operations	
Review the four symbols students have learned so far, using number bonds. Write the expressions and have students draw number bonds for each and find the answer. They can use counters for the multiplication and division problems. Point out that with subtraction and division we start with the total. With addition and multiplication, we want to find the total. In addition and subtraction, the parts do not have to be the same. We can add more than two parts, but they do not have to be equal. In multiplication and division, all the parts have to be the same.	8 + 2 = 10 8 x 2 = 16 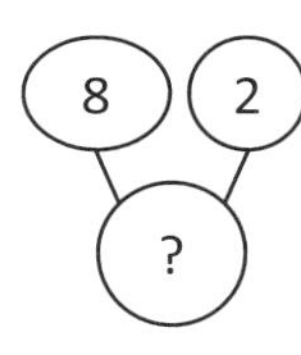 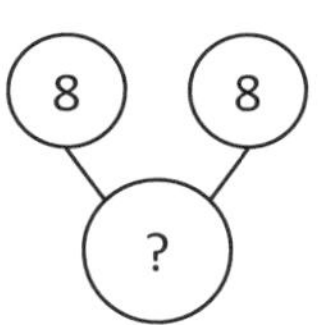 8 – 2 = 6 8 ÷ 2 = 4 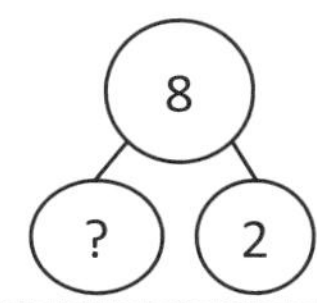 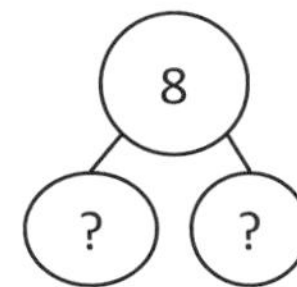
Assessment	**Text p. 85**
	7. 9 x 2 = **18** 2 x 9 = **18** 18 ÷ 2 = **9** 18 ÷ 9 = **2** 8. 8 x 4 = **32** 4 x 8 = **32** 32 ÷ 4 = **8** 32 ÷ 8 = **4**
Write several addition, subtraction, multiplication, and division problems on the board and have students copy them and write the answers. They can use counters or multilink cubes to solve the multiplication and division problems.	10 + 2 (12) 10 – 2 (8) 10 x 2 (20) 10 ÷ 2 (5) 24 – 6 (18) 24 ÷ 6 (4) 24 + 6 (30) 9 + 3 (12) 9 x 3 (27)
Practice	WB Exercise 39, pp. 107-109

Exercise 39

1. (a) 6
   6

   (b) 5
   5

107

2. (a) 3 2

   (b) 7 3

   (c) 5 4

   (d) 9 2

108

3. (a) 35 ÷ 7 = 5   35 ÷ 5 = 7

   (b) 18 ÷ 6 = 3   18 ÷ 3 = 6

4. 6 x 4 = 24   4 x 6 = 24
   24 ÷ 6 = 4   24 ÷ 4 = 6

109

5.2d Practice and Review

Objectives

- Solve word problems which involve multiplication and division.
- Review all topics.

Note

You may want to spend several class periods on this lesson.

In the workbook review, students are supplied with pictures for the multiplication and division problems. For the addition and subtraction problems, they can draw number bonds if needed, but may not need them by now.

Practice	**Text pp. 86-87, Practices 5B and 5C**
	Practice 5B 1. 12 ÷ 2 = **6** Each child gets 6 oranges. 2. 24 ÷ 6 = **4** There are 4 boxes. 3. 30 ÷ 3 = **10** There were 10 sticks in each bundle. 4. 28 ÷ 4 = **7** She needs 7 boxes. Practice 5C 1. 7 x 3 = **21** 3 x 7 = **21** 21 ÷ 3 = **7** 21 ÷ 7 = **3** 2. 18 ÷ 6 = **3** There were 3 bags of pears. 3. 5 x 4 = **20** He bought 20 books. 4. 35 ÷ 5 = **7** Each child gets 7 cookies.
For additional practice, you can have students to the problems on appendix p. a30. Allow students to use counters or drawings to solve the problems.	Appendix p. a30
1. Harriet has 16 cookies and wants to divide them up among herself and three friends. How many would each person get?	16 ÷ 4 = 4 Each person would get 4 cookies.
2. Harriet has 20 cookies. She wants to put 4 cookies on each plate. How many plates does she need?	20 ÷ 4 = 5 She needs 5 plates.
3. There are 6 plates of cookies with 5 cookies on each plate. How many cookies are there?	6 x 5 = 30 There are 30 cookies in all.

4. On a rainy day, Harriet's friends all wore boots when they came to visit. They left 12 boots by the door. How many friends came to visit?	12 ÷ 2 = 6 6 friends came to visit.
5. You see some dogs and count 32 legs. How many dogs are there?	32 ÷ 4 = 8 There are 8 dogs.
6. Harriet uses 14 jugs of water to give water to her birds in a week. If she used the same number of jugs each day, how many jugs did she use each day?	14 ÷ 7 = 2 She used 2 jugs a day.
7. A string 12 inches long is cut into equal parts, each 3 inches long. How many equal parts are there?	12 in. ÷ 3 in. = 4 There are 4 equal parts.
8. A string 12 inches long is cut into two equal parts. How long is each part?	12 in. ÷ 2 = 6 in. Each part is 6 inches long.
9. There are 3 feet in a yard and 12 inches in a foot. How many inches are there in a yard?	3 x 12 in. = 36 in. There are 36 inches in a yard.
10. A decimeter is 10 centimeters. How many decimeters are there in a meter? (Let students use a meter stick and count how many groups of 10 cm there are.)	100 cm ÷ 10 cm per dm = 10 dm There are 10 decimeters in a meter.
Review	WB Review 5, pp. 110-113

3. (a)

750	740	**730**	720	**710**	700	**690**

(b)

392	**492**	592	**692**	792	**892**	992

Review 5

1. 24 + 17 = 41 17 + 24 = 41
   41 - 17 = 24 41 - 24 = 17

2. 8 x 2 = 16 2 x 8 = 16
   16 ÷ 2 = 8 16 ÷ 8 = 2

3. (a) The numbers decrease by 10. See table above.
   (b) The numbers increase by 100. See table above.

110

4. (a) 465
   (b) 261
   (c) 397
   (d) 742
   (e) 44
   (f) 665
   (g) 738
   (h) 199

5. (a) 957
   (b) 980
   (c) 980
   (d) 957

111

6. 20 ÷ 5 = 4
   4

7. 18 ÷ 6 = 3
   3

8. 5 x 4 = 20
   20

112

9. 124 - 48 = 76
   76

10. 98 cm + 14 cm = 112 cm
    112 cm

11. 290 g - 132 g = 158 g
    158 g

113

6 Multiplication Tables of 2 and 3

Objectives

- Count by twos and threes.
- Build multiplication tables for 2 and 3.
- Practice multiplication facts for 2 and 3.
- Use multiplication facts for 2 and 3 to divide by 2 and 3.
- Practice division facts for 2 and 3.
- Solve word problems involving multiplication or division.

Suggested number of weeks: 3-4

		TB: Textbook WB: Workbook	Objectives	Material	Appendix
6.1	**Multiplication Table of 2**				
6.1a	Count by Twos	TB: pp. 88-90 WB: pp. 114-119	♦ Count by twos. ♦ Solve problems in the form 2 x ___ = ___ by counting by twos.	♦ Multilink cubes	
6.1b	Multiplication Table of 2	TB: pp. 90-91 WB: pp. 120-122	♦ Build the multiplication table for 2. ♦ Compute unknown facts from known facts. ♦ Practice multiplication facts for 2.	♦ Multilink cubes ♦ Number cards	Mental Math 19 a31
6.1c	Doubles	TB: pp. 91-92 WB: pp. 123-124	♦ Relate the facts for ___ x 2 to the facts for 2 x ___. ♦ Practice multiplication facts for 2.	♦ Multilink cubes ♦ Number strips ♦ Counters ♦ Number cards	Mental Math 20-21 a32-a33
6.1d	Practice	TB: p. 93 WB: pp. 125-128	♦ Practice multiplication facts for 2. ♦ Solve word problems.		
6.2	**Multiplication Table of 3**				
6.2a	Count by Threes	TB: pp. 94-95 WB: pp. 129-133	♦ Count by threes. ♦ Solve problems in the form 3 x ___ = ___ by counting by threes.	♦ Hundred-chart ♦ Multilink cubes	
6.2b	Triples	TB: p. 96 WB: pp. 134-135	♦ Relate the facts for ___ x 3 to the facts for 3 x ___. ♦ Practice multiplication facts for 3.	♦ Multilink cubes ♦ Fact cards with answer cards	a34

		TB: Textbook WB: Workbook	Objectives	Material	Appendix
6.2c	Multiplication Table of 3	TB: pp. 96-97 WB: pp. 138-141	♦ Build the multiplication table for 3. ♦ Compute unknown facts from known facts. ♦ Practice multiplication facts for 3.	♦ Multilink cubes ♦ Number cards ♦ Fact cards	Mental Math 22-23 a35-a36
6.2d	Practice	TB: p. 98 WB: pp. 136-137, 142-144	♦ Practice multiplication facts for 3. ♦ Solve word problems.	♦ Fact cards	Mental Math 24
6.2e	Practice	TB: p. 99 WB: pp. 145-147	♦ Practice multiplication facts for 2 and 3. ♦ Solve word problems.	♦ Number cards ♦ Number cubes ♦ Counters	Mental Math 25-27 a37
6.3	**Dividing by 2**				
6.3a	Divide by 2	TB: pp. 100-101 WB: pp. 148-149	♦ Relate division by 2 to multiplication by 2.	♦ Counters	a38
6.3b	Practice	TB: pp. 102-103 WB: pp. 150-151	♦ Practice division facts for 2. ♦ Solve word problems.	♦ Number cards	Mental Math 28-29 a39
6.3c	Practice	TB: p. 104	♦ Practice multiplication and division facts for 2. ♦ Solve word problems.	♦ Counters	Mental Math 30 a40
6.4	**Dividing by 3**				
6.4a	Divide by 3	TB: p. 105 WB: pp. 152-153	♦ Relate division by 3 to multiplication by 3.	♦ Counters ♦ Fact cards	Mental Math 31-32 a41-a42
6.4b	Practice	TB: pp. 106-107 WB: pp. 154-157	♦ Practice division facts for 2 and 3. ♦ Solve word problems.	♦ Fact cards with answer cards	Mental Math 33
6.4c	Practice	TB: p. 108 WB: pp. 158-162	♦ Practice multiplication and division facts for 2 and 3. ♦ Solve word problems.		Mental Math 34-36
6.4d	Review	TB: pp. 109-112 WB: pp. 163-174	♦ Review all previous topics.		

6.1 6.2 Multiplication Tables of 2 and 3

Objectives

- Count by twos and threes.
- Build multiplication tables for 2 and 3.
- Practice multiplication facts for 2 and 3.
- Solve word problems.

Material

- Counters
- Multilink cubes
- Fact cards
- Number cards 1-30
- Number cards 1-10, 4 sets per group
- Number strips
- Hundred chart
- Mental Math 19-27
- Appendix pp. a31-a37

Prerequisites

Initially, students will use fingers to keep track of how many twos they have counted by in order to find the answer to a multiplication problem. They need to be able to know by sight how many fingers are being held up so that they don't get confused when they get to the answer, as in knowing that it is 7 fingers (5 on one hand and 2 on the other) when they count by twos to find 2 x 7. Students did learn to count by twos in *Primary Mathematics* 1 so you may not have to spend much time on counting by twos.

Students need to understand the concepts in the previous unit: that multiplication involves equal groups and that we can find the answer to multiplication problems by adding the amounts in the equal groups.

They should be able to add 2 mentally.

Notes

In this unit, students will begin to study and commit multiplication facts to memory. In Part 1 they will learn the multiplication facts for 2. In Part 2 they will learn the multiplication facts for 3.

By the end of this unit students should know most of the facts for 2 and 3 for both multiplication and division, or be able to quickly calculate the ones they haven't memorized using various strategies involving mental math. Students will first be given skills to help them calculate a fact (e.g., counting by twos or threes, using a known fact to find an unknown fact) and then will begin memorizing facts. They will then have the means to quickly figure out a fact they might forget later.

Some students who are quite good at math concepts have difficulty in memorizing math facts. They can become quite fast at calculating them. A student is not necessarily a poor math student just because he or she does not have instant recall of math facts; speed with math facts is not equivalent to mathematical ability, but simply helps with speed. Continue to have students work on any troublesome facts a little each day, without letting memorization become burdensome. Use sprints to add some fun and individual accomplishment.

You may wish to include 2 x 11 and 2 x 12, since students in the U.S. will encounter twelves frequently (e.g., 12 inches in a foot). However, multiplication of a 2-digit number by a 1-digit number (e.g., 12 x 2) will be covered in *Primary Mathematics* 3, and students will learn to multiply a 2-digit number, particularly easy ones like 11 and 12, by a 1-digit number mentally.

Even and odd numbers will be taught along with division with a remainder in *Primary Mathematics* 3.

Students learned how to count by twos, fives, and tens previously. In this unit they will also learn to count by threes. Counting by the other numbers — 4s, 6s, 7s, 8s, and 9s — as a way of memorizing math facts is not emphasized in this curriculum since students are given other tools to quickly calculate a fact they do not know. Once they have completed *Primary Mathematics* 2B, students will know all but 16 of the multiplication facts through 10 x 10 and the related division facts. The last 16 are facts where both the factors are only 6, 7, 8, or 9. If they know the fact for 3 x 8, then they also know the fact for 8 x 3. When they work on learning the facts for 6, 7, 8, and 9 in *Primary Mathematics* 3A, they will only need to concentrate on the last 16 facts and will be given strategies to find them from known facts, rather than counting by those numbers.

6.1a Count by Twos

Objectives

- Count by twos.
- Solve problems in the form 2 x ___ = ___ by counting by twos.

Note

In this lesson, students will relate counting by twos to multiplication by 2. In order to keep track when counting by twos to solve a multiplication problem initially, they can use their fingers.

More capable students will soon be able to do these problems mentally by counting by twos. Less capable students can use manipulatives by selecting the appropriate number of two-linked cubes and counting.

Count by twos	
Write the numbers 1 through 20 in order on the board and get students to read them. Tell them they are counting by ones. Each number is 1 more than the last number. Now, point to the 2, then the 4, and so on and have students read the numbers as you do so. Tell them they are now counting by twos since each number is two more than the last number they counted. Erase the odd numbers and have them read the even numbers.	1 2 3 4 5 6 7 8 9 10 11 12 13 14 15 16 17 18 19 20 2 4 6 8 10 12 14 16 18 20
Display or draw one set of 2-linked cubes or squares. Write 2 under it. Then display another set and ask how many there are altogether. Write 4 under the second set. Continue to 20. Then get students to read the numbers again in order.	2 4 6 8 10 12 14 16 18 20
Have students practice counting by twos, both forward and backward. It is likely that they can count by twos to 20 easily. You can extend counting by twos past 20 in future practice.	
Say or write a number between 1 and 20 and ask students if it is in the counting by twos sequence. For example, 15 is not, but 18 is.	
Multiply 2	
Set out 5 sets of 2-linked cubes. Have students count by twos to find how many there are. Write the addition equation. Tell students that since we have equal groups of 2, we are multiplying 2 by 5, so we can also write the multiplication equation: 2 x 5 = 10. Display another set of equal groups of 2 and ask students for a multiplication expression and the answer.	2 + 2 + 2 + 2 + 2 = 10 2 x 5 = 10
Discussion	**Text pp. 88-89**
Show students how they can hold up one finger for each two to help them keep track of how many twos they have counted. For example, with 2 x 7, they hold up one finger as they say 2, then another as they say 4, and so on until they have 7 fingers up, at which point they should say 14, which is the answer to 2 x 7.	(a) 6; 6 (b) 14; 14

Assessment	**Text p. 90**
	1. (a) 4 (b) 18
Give students a few word problems involving multiplication of 2 2 x ___) and have them write the expression and find the answer by counting by twos. They may use 2-linked cubes if needed. (Memorization of the facts is not required yet).	
⇒ How many feet are there on 4 parrots?	2 x 4 = 8; There are 8 feet.
⇒ How many wings are there on 6 macaws?	2 x 6 = 12; There are 12 wings.
⇒ 2 cookies each were given out to 10 children. How many cookies were given out?	2 x 10 = 20; 20 cookies were given out.
Write some 2 x ___ multiplication problems and have students work out the answers by counting by twos. Include 2 x 1. You can also include 2 x 11 and 2 x 12.	2 x 1 (1) 2 x 9 (18) 2 x 11 (22)
Practice	WB Exercise 40, pp. 114-117 WB Exercise 41, pp. 118-119

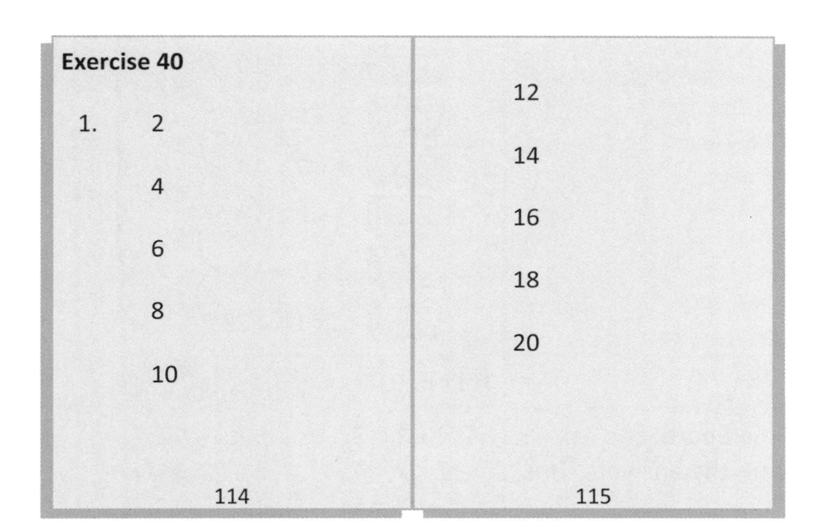
Exercise 40

1. 2
4
6
8
10

114

12
14
16
18
20

115

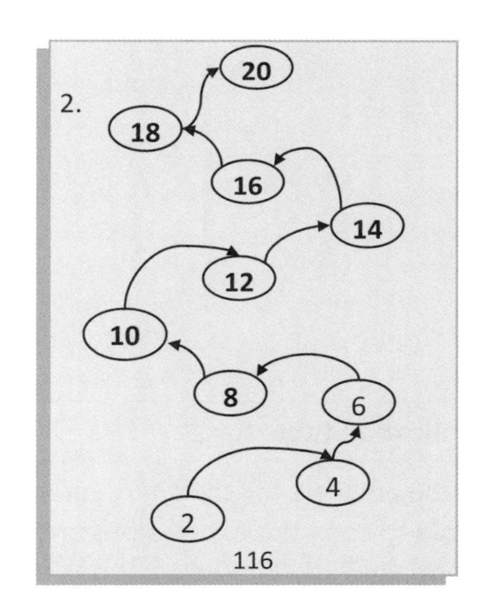

3. 4
6
8
10
12
14
16
18
20

117

Exercise 41

1. 6
6

2. 8
8

3. 10
10

118

4. 12
12

5. 2 x 7 = 14
14

6. 2 x 8 = 16
16

119

6.1b Multiplication Table of 2

Objectives

- Build the multiplication table for 2.
- Compute unknown facts from known facts.
- Practice multiplication facts for 2.

Note

Although students will need to commit multiplication facts to memory, as they do so they need strategies to help them compute the ones they forget. An unknown fact can be computed from a known fact by adding on twos or subtracting twos. This idea is taught in this lesson, and will be used throughout for multiplication by other numbers than 2. Students who have difficulties memorizing isolated facts but are otherwise quite able to grasp mathematical concepts will make more use of these strategies. The strategies also help gain a better understanding of multiplication.

Two more and two less	
Set out a 2-linked cube or draw it on the board and write the multiplication sentence next to it. Continue adding 2-linked cubes and writing the multiplication sentences. Stress the idea that the next one is "2 more." You can draw an arrow from one row to the next with +2 next to it. Continue to 2 x 10 = 20. Then, read the answers from the bottom up, stressing that each one is "2 less."	2 x 1 = 2 2 x 2 = 4 2 x 3 = 6 2 x 4 = 8 2 x 5 = 10 2 x 6 = 12 2 x 7 = 14 2 x 8 = 16 2 x 9 = 18 2 x 10 = 20 (+2 between each row going down; −2 between each row going up)
Multiplication table for 2	**Text p. 91**
Erase the answers for the chart you have on the board, and ask students to copy the expressions and then write the answers. This is essentially Task 4 in the textbook on p. 91.	4. 2 x 1 = 2 2 x 6 = **12** 2 x 2 = 4 2 x 7 = **14** 2 x 3 = 6 2 x 8 = **16** 2 x 4 = **8** 2 x 9 = **18** 2 x 5 = **10** 2 x 10 = **20**
Find an unknown fact from a known fact.	
Find a fact students have already memorized just from using it, such as 2 x 5. Write the equation on the board, and set out the corresponding number of 2-linked cubes. Point out that if we add 2 more cubes, then will have 2 x 6 cubes. So if we know the answer to 2 x 5, to find the answer to 2 x 6 we can count up 2 from the answer to 2 x 5, rather than counting by twos all the way from the beginning. Repeat with some other examples.	2 x 5 = 10 10 2 x 6 = 10 + 2 = 12

Assessment	Text p. 90
	2. (a) 6 (b) 8 3. 12
Multiplication facts	
Start working on helping students to commit the multiplication facts for 2 to memory. Use number cards 1-10. Mix them up and hold them up one at a time and see how fast students can find 2 multiplied by that number. Go as quickly as you can. Some students won't answer, but they can start memorizing by hearing the answers. If time permits, work with individual students. Students can work individually with a deck of number cards, turning them over one at a time and giving the fact for 2 multiplied by the number that is turned over.	Number cards 0-10
Start working on Sprints (seeing how many problems can be answered in 1 minute) for multiplication facts for 2 as you proceed.	Mental Math 19
Practice	WB Exercise 42, pp. 120-121 WB Exercise 43, p. 122

Exercise 42

1. (a) 8

(b) 12

(c) 14
16

120

2. 6
8
10
12
14
16
18
20

121

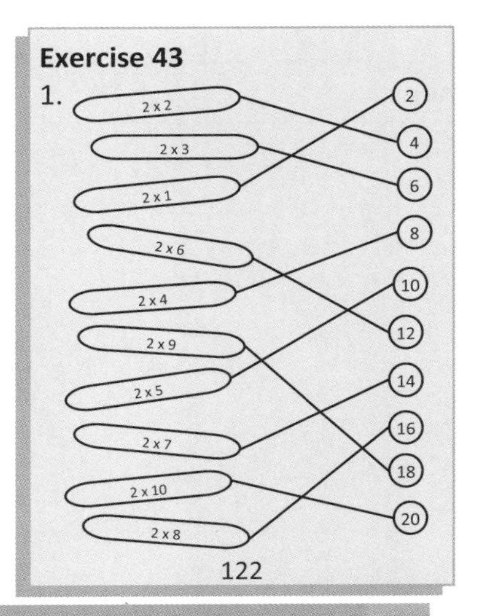

6.1c Doubles

Objectives

- Relate the facts for ___ x 2 to the facts for 2 x ___.
- Practice multiplication facts for 2.

Note

Although a chart is provided for Tasks 4 and 7 on pages 91 and 92 (appendix p. a31), it may help some students to remember the facts if they write the entire equation. Some students may benefit from hearing or reading the entire equations out loud.

Arrays	
Use multilink cubes to create a 5 by 2 array. Split it apart into twos and write or have students write the addition equation and 2 multiplied by 5 equation. Put the array back together and split apart into fives. Write 5 + 5 = 10 and then 5 x 2 = 10. Point out that we have the same total, and so the answer is the same, whether we multiply 2 by 5 or 5 by 2. Point out that multiplying by 2 is the same as adding a number to itself, or doubling the number. We can find the answer to both 2 x 5 and 5 x 2 by finding the answer to 5 + 5.	2 + 2 + 2 + 2 + 2 = 10 2 x 5 = 10 5 + 5 = 10 5 x 2 = 10 2 x 5 = 5 x 2
Multiplication table	**Text p. 92** (appendix p. a31)
Have students fill out a copy of the chart on p. a31 in the appendix, or do Task 7 on p. 92 of the textbook, copying the equations.	7. 1 x 2 = 2 6 x 2 = **12** 2 x 2 = 4 7 x 2 = **14** 3 x 2 = 6 8 x 2 = **16** 4 x 2 = **8** 9 x 2 = **18** 5 x 2 = **10** 10 x 2 = **20**
Assessment	**Text pp. 91-92**
	5. (a) 5 + 5 = **10** (b) 7 + 7 = **14** 5 x 2 = **10** 7 x 2 = **14** 6. (a) 8 + 8 = **16** (b) 9 + 9 = **18** 8 x 2 = **16** 9 x 2 = **18** 8. 6 x 2 **2** x 7 2 x 9 2 x **6** 7 x 2 **9** x 2 9. 6 x 2 = **12** The total length was **12** m.

Multiplication facts	Mental Math 20-21
Show students how to fill out a copy of the partial multiplication chart on p. a32 of the appendix. Then have them fill out copies of charts on p. a33 now or later.	Appendix pp. a32-33
Practice	WB Exercise 44, pp. 123-124

Group Game

Material: Strips containing the even numbers through 20 and 10 counters for each student. Number cards 1-10, 4 sets per group.

Procedure: The cards are shuffled and placed face down in the middle. The players take turns drawing a card. They double the number on the card and cover the answer on their number strip (if it is not already covered). The first player to get all the numbers covered wins.

2	4	6		10	12	14	16	18	20

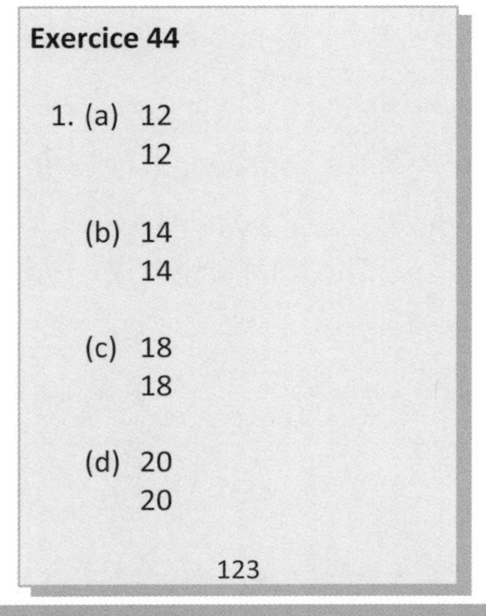

Exercice 44

1. (a) 12
   12

   (b) 14
   14

   (c) 18
   18

   (d) 20
   20

123

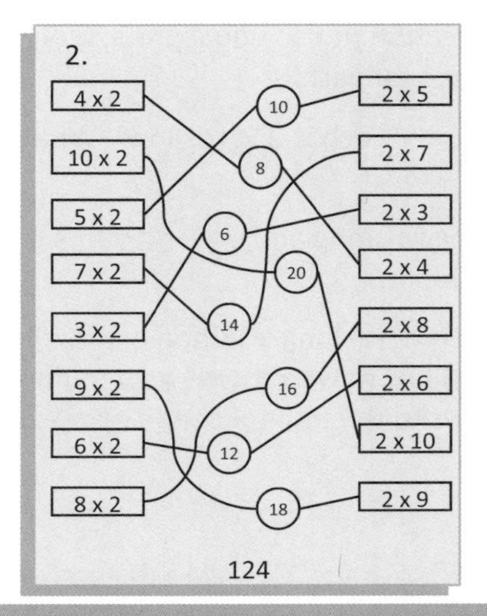

6.1d Practice

Objectives

- Practice multiplication facts for 2.
- Solve word problems.

Note

You may want to take several days for this lesson.

Students should attempt to answer multiplication problems for twos from memory as much as possible. Continue to provide fact practice as you proceed.

Practice	**Text p. 93, Practice 6A**
Have students solve the word problems and then discuss their solutions. They should write multiplication problems for each of the word problems. If needed, they can draw simple diagrams to help them interpret the problems, but do not turn this into a drawing class. For Task 6, they can draw simple birds with wings, or even better, circles with 2 in each to represent the wings. Discuss how many groups there are and what is in each group for selected problems. In problems 6, 7, and 9, there are groups of 2. In problem 8 there are 2 groups of \$5. In problem 10, there are 2 groups of 8 m.	1. (a) 6 (b) 8 (c) 4 2. (a) 2 (b) 18 (c) 16 3. (a) 12 (b) 14 (c) 20 4. (a) 10 (b) 6 (c) 8 5. (a) 18 (b) 12 (c) 14 6. (2) (2) (2) (2) (2) (2) 2 x 6 = **12** or 6 x 2 = 12 6 birds have 12 wings. 7. 2 x 10 = **20** 10 children get 20 balloons. 8. \$5 x 2 = **\$10** He can save \$10 in 2 weeks. 9. 4 x 2 kg = **8 kg** She bought 8 kg of coffee. 10. [8 m] [8 m] 2 x 8 m = **16 m** She used 16 m of cloth.
Have students solve the following problems and discuss why one is solved with multiplication (it has equal groups) and the other with addition (it has 2 unequal parts). ⇒ There are 8 ribbons. Each is 2 yards long. What is the total length of ribbon? ⇒ One ribbon is 8 yards long and one is 2 yards long. What is the total length of ribbon? Point out that we can write long addition expressions even when we have equal parts, but now that we know multiplication facts, it is much quicker to write the multiplication expression.	 8 x 2 yd = 16 yd The total length is 16 yd. 8 yd + 2 yd = 10 yd The total length is 10 yd.

Multiplication facts	
Continue to provide practice with the multiplication facts. You can use the games suggested earlier, the mental math pages, or other activities.	
Enrichment	
Write the expression 2 x 5 and ask students for the answer. Tell them that 2 x 5 is the same as 10, so 2 x 5 = 10 is a true statement. Ask students for suggestions for other expressions that 2 x 5 could equal. Write some of them down. Write some others using 5 + 5 and 5 x 2.	2 x 5 = ___ 2 x 5 = 10 2 x 5 = 5 + 5 2 x 5 = 2 + 2 + 2 + 2 + 2 2 x 5 = 8 + 2 2 x 5 = 12 – 2 2 x 5 = 5 x 2 5 + 5 = 8 + 2 5 x 2 = 12 – 2
Write the equation 10 = 9 and ask students if it is true. 10 is not the same as 9, so this is a false statement. Write 2 x 5 = 5 + 4 and ask students if it is true. It is false, since 2 x 5 is 10, but 5 + 4 is 9. So the expressions on each side of the equal sign are not equal, and the whole statement is false. Point out that since we know that 2 x 5 = 5 + 5, it can't equal 4 + 5.	10 = 9 ? 2 x 5 = 5 + 4 ?
Write a few more examples, such as those shown here, some false and some true, and ask students if they are true or not. With 6 x 8 = 8 x 6, students do not need to evaluate each side to determine if it is true or not.	5 + 3 = 3 + 5 ? (True) 5 + 3 = 3 + 2 ? (False) 6 x 2 = 2 + 2 + 2 + 2 + 2 ? (False) 2 x 6 = 10 + 6 ? (False) 2 x 2 = 2 + 2 ? (True) 2 x 7 = 12 + 2 ? (True) 2 x 10 = 5 + 5 ? (False) 6 x 8 = 8 x 6 ? (True)
Practice	WB Exercise 45, pp. 125-126 WB Exercise 46, pp. 127-128

Exercice 45

1.
8
20
10
16
4
6
14
12
18
2

125

2.

6	16	12
8	18	14
20	10	8
16	18	12

126

Exercice 46

1. 16
   16

2. 6 x \$2 = \$12
   \$12

3. 3 x 2 = 6
   6

127

4. 2 x \$7 = \$14
   \$14

5. 5 x 2 kg = 10 kg
   10 kg

6. 2 m x 4 = 8 m
   8 m

128

6.2a Count by Threes

Objectives

- Count by threes.
- Solve problems in the form 3 x ___ = ___ by counting by threes.

Note

Although students have not specifically learned to count by threes in the *Primary Mathematics* curriculum, they have often added three by counting on or using mental math strategies. In this lesson, they will practice counting by threes so that they become more familiar with multiples of 3 and can find answers to problems involving multiplication of 3 as they start memorizing the multiplication facts for 3.

<table>
<tr><td>Count by threes</td><td></td></tr>
<tr><td>Write the numbers 1 through 30 in order, or use the first three rows of a hundred-chart. Circle every third number. You can go beyond 30 if you want. Students can do the same with individual hundred-charts. Ask students to read each marked number. Ask them what they are counting by. They are counting by threes, since each number is three more than the previous circled number.</td><td>
<table>
<tr><td>1</td><td>2</td><td>3</td><td>4</td><td>5</td><td>6</td><td>7</td><td>8</td><td>9</td><td>10</td></tr>
<tr><td>11</td><td>12</td><td>13</td><td>14</td><td>15</td><td>16</td><td>17</td><td>18</td><td>19</td><td>20</td></tr>
<tr><td>21</td><td>22</td><td>23</td><td>24</td><td>25</td><td>26</td><td>27</td><td>28</td><td>29</td><td>30</td></tr>
<tr><td>31</td><td>32</td><td>33</td><td>34</td><td>35</td><td>36</td><td>37</td><td>38</td><td>39</td><td>40</td></tr>
</table>
</td></tr>
<tr><td>Display a 3-linked cube or draw it on the board and ask students how many there are. Write 3 under it. Then show another 3-linked set and ask how many there are altogether. Write 6 under it. Continue to 30 (or 36 if you are including 11 and 12). Be sure students understand that you are writing the total number up to that point. Then remove the cubes and have students read the numbers.</td><td>3 6 9 12 15 18 21 24 27 30</td></tr>
<tr><td>Have students practice counting by threes.</td><td></td></tr>
<tr><td>Say or write a number between 1 and 30 and ask students if it is in the counting by threes sequence. For example, 20 is not, but 18 is.</td><td></td></tr>
<tr><td>Multiply 3</td><td></td></tr>
<tr><td>Set out 5 sets of 3-linked cubes. Have students count by threes to find how many there are. Write the addition equation. Tell them that since we have equal groups of 3, we are multiplying 3 by 5. Ask them for the multiplication equation.

Display another set of 3-linked cubes and ask students for the multiplication expression and the answer.</td><td>3 + 3 + 3 + 3 + 3 = 15
3 x 5 = 15</td></tr>
<tr><td>Discussion</td><td>Text pp. 94-95</td></tr>
<tr><td></td><td>(a) 15; 15
(b) 27; 27</td></tr>
</table>

Assessment	**Text p. 95**
	1. (a) 12 (b) 24
Give students a few word problems involving multiplication of 3 and have them write the expression and find the answer by counting by threes. They may use 3-linked cubes if needed.	
⇒ How many wheels are there on 7 tricycles?	3 x 7 = 21; There are 21 wheels.
⇒ A pound of birdseed costs \$3. How much do 4 pounds cost?	\$3 x 4 = \$12; 4 pounds costs \$12.
⇒ 3 cookies each were given out to 10 children. How many cookies were given out?	3 x 10 = 30; 30 cookies were given out.
Reinforcement	
Write some 3 x ___ multiplication problems and have students work out the answers by counting by threes. Include 3 x 1. You can also include 3 x 11 and 3 x 12.	3 x 1 (3) 3 x 9 (27) 3 x 11 (33)
Practice	WB Exercise 47, pp. 129-131 WB Exercise 48, pp. 132-133

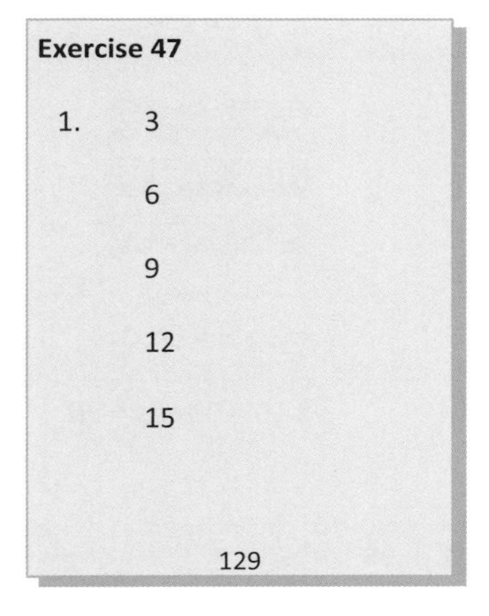

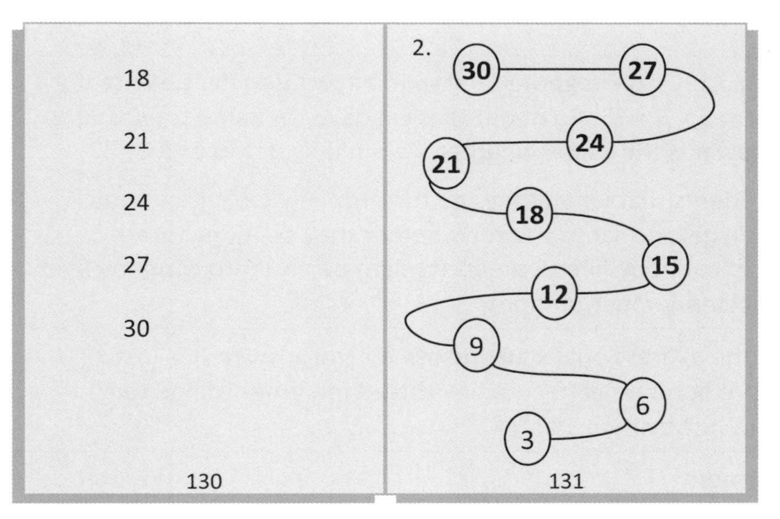

Exercise 48

1. 6
9
12
15
18
21
24
27
30

132

2. (a)

1	2	4	6	8
2	**4**	**8**	**12**	**16**

(b)

1	2	5	7	9
3	**6**	**15**	**21**	**27**

(c)

3	**\$6**
5	**\$10**
7	**\$14**
9	**\$18**
10	**\$20**

3	**\$9**
4	**\$12**
6	**\$18**
8	**\$29**
10	**\$30**

133

6.2b Triples

Objectives

- Relate the facts for ___ x 3 to the facts for 3 x ___.
- Practice multiplication facts for 3.

Note

This lesson could be combined with the next one.

Arrays	
Use multilink cubes to create a 5 by 3 array. Split it apart into threes and write the expression **3 x 5**, and then have students supply the answer.	3 x 5 = 15
Put the array back together and split it apart into fives. Write the equation **5 x 3 = 15**. Point out that we have the same total, and so the answer is the same, whether we multiply 3 by 5 or 5 by 3. Tell students that when they see the problem 3 x 5, they know they can get the same answer whether they count by threes 5 times or count by fives 3 times. It might be easier to count by fives the shorter number of times.	5 x 3 = 15 3 x 5 = 5 x 3
Write the expression **3 x 10** and ask for the answer. It is just 3 tens. We don't need to count by threes ten times to find the answer, just by tens 3 times.	3 x 10 = 3 tens = 30
Assessment	**Text p. 96**
	2. 3 x 6 = **18** 6 x 3 = **18** 3. 2 x 3 **3** x 7 3 x 8 3 x **2** 7 x 3 **8** x 3
Ask students to draw 3 long lines going one way and then 6 going a different way so that they cross the 3 lines. Ask them to write two multiplication equations for the number of intersections, or places where the lines cross each other.	6 x 3 = 18 3 x 6 = 18

Multiplication facts	Appendix p. a34
Show students how to fill out a copy of the partial multiplication chart on p. a34 of the appendix.	
Practice	WB Exercise 49, pp. 134-135

Group Game

Material: A set of fact cards for 2 x ___, ___ x 2, 3 x ___, and ___ x 3, without answers on the back, and one set of answers. Cards should be shuffled.

Procedure: The dealer deals 5 cards to each player and puts the rest face-down in the center. Players take turns asking another student for a particular card, "Do you have two times three?" If the other player does not have a card for the requested fact, the player whose turn it is draws a card from the center pile. If the other player does have the card, he must give it to the player who asked for it. The goal is to make sets of three containing both fact cards and the answer (e.g., 2 x 3, 3 x 2, and 6). Completed sets are put face-up in front of the player.

Exercise 49

1. 6
   6

2. (a) 9

   (b) 12
   12

   (c) 15
   15

134

3. (a) 18
   18

   (b) 21
   21

   (c) 24
   24

   (d) 27
   27

135

6.2c Multiplication Table of 3

Objectives

- Build the multiplication table for 3.
- Compute unknown facts from known facts.
- Practice multiplication facts for 3.

Note

Exercises 51 and 52 are assigned with this lesson. Exercise 50 is assigned with the next lesson, after students have practiced multiplication facts more.

Three more and three less	
Set out a 3-linked cube or draw it on the board and write the related multiplication sentence next to it. Continue adding 3-linked cubes and writing the multiplication sentences. Stress the idea that the next one is "3 more." You can include both related multiplication expressions for each line. Point out that 3 x 3 is the same both ways. Then have students read the answers from the bottom up, stressing that each one is "3 less."	3 x 1 = 1 x 3 = 3 3 x 2 = 2 x 3 = 6 3 x 3 = 3 x 3 = 9 3 x 4 = 4 x 3 = 12 3 x 5 = 5 x 3 = 15 3 x 6 = 6 x 3 = 18 3 x 7 = 7 x 3 = 21 3 x 8 = 8 x 3 = 24 3 x 9 = 9 x 3 = 27 3 x 10 = 10 x 3 = 30 (+3 between each line going down; −3 going up)
Multiplication Table	**Text p. 96** (appendix p. a35)
Erase the answers for the chart you have on the board, and ask students to copy the expressions and then write the answers. This is essentially Task 4 in the textbook on p. 96. You can also copy the chart on appendix p. a35 and have students fill it in.	4. 3 x 1 = 3 1 x 3 = **3** 3 x 2 = 6 2 x 3 = **6** 3 x 3 = 9 3 x 3 = **9** 3 x 4 = **12** 4 x 3 = **12** 3 x 5 = **15** 5 x 3 = **15** 3 x 6 = **18** 6 x 3 = **18** 3 x 7 = **21** 7 x 3 = **21** 3 x 8 = **24** 8 x 3 = **24** 3 x 9 = **27** 9 x 3 = **27** 3 x 10 = **30** 10 x 3 = **30**

Find an unknown fact from a known fact	
Write the expression **3 x 10** on the board and ask students for the answer. Then write the expression **3 x 9**. Tell students that it is easy to find 3 x 10 since it is just 3 tens. It is harder to remember what 3 x 9 is. Ask them how we can use the answer from 3 x 10 to get the answer to 3 x 9. We can subtract 3 from 30. Write the expression **3 x 11** on the board and ask how we can find the answer using the fact for 3 x 10. We can add 3. If necessary, use 3-linked cubes or other manipulatives to illustrate.	3 x 10 = 30 3 x 9 = 30 – 3 = 27 3 x 11 = 30 + 3 = 33
Assessment	**Text p. 97**
	5. 18 6. 24 7. 7 x 3 = **21** 21 kg
Multiplication facts	Mental Math 22-23
Continue helping students memorize the multiplication facts for 3. You can use adaptations of earlier activities.	
Have them fill out copies of charts on p. a36 now or later.	Appendix p. a36
Practice	WB Exercise 51, pp. 138-139 WB Exercise 52, pp. 140-141

Game

Material: Cards with multiples of 2 or multiples of 3, five cards per student. Large cards with the multiplication expressions for multiplying by twos or threes.

Procedure: Give each student 5 cards. They place their cards in front of them face-up. Hold up one card at a time. If a student has a card containing the answer, he or she turns that card over (face-down). The goal is for each student to have all of his or her cards turned over. If you are playing this game with this particular lesson, allow students time to find the answers to the __ x 3 or 3 x __ cards by counting by threes as needed since they have not yet had much practice memorizing these facts yet.

Exercise 51

1. (a) 12
   (b) 18
   (c) 24
   27

138

2. 9
12
15
18
21
24
27
30

139

Exercise 52

1. (a) 12
   (b) 18
   (c) 30
   27

140

2. 24
21
18
15
12
9
6
3

141

6.2d Practice

Objectives

- Practice multiplication facts for 3.
- Solve word problems.

Note

You may want to take several days for this lesson in order to spend more time practicing multiplication facts.

Students should attempt to answer multiplication problems for threes from memory as much as possible. Continue to provide fact practice as you proceed.

Practice	**Text p. 98, Practice 6B**
Have students solve the word problems and then discuss their solutions. They should write multiplication problems for each of the word problems. If needed, they can draw simple diagrams to help them interpret the problems. Discuss how many groups there are and what is in each group for selected problems. In problems 6, 9, and 10, there are groups of 3. In problem 7 there are 3 groups of 7 trees. In problem 8, there are 3 groups of 8 kg.	1. (a) 3 (b) 6 (c) 12 2. (a) 18 (b) 21 (c) 24 3. (a) 12 (b) 15 (c) 30 4. (a) 21 (b) 27 (c) 9 5. (a) 24 (b) 30 (c) 18 6. 3 x 4 = **12** There are 12 wheels on 4 tricycles. 7. 7 x 3 = **21** There are 21 trees in 3 rows. 8. 3 x 8 kg = **24 kg** 3 bags of potatoes weigh 24 kg. 9. 6 x 3 yd = **18 yd** She used 18 yd of cloth. 10. 3 x 10 = **30** He bought 30 stamps.
Write a multiplication expression for multiplication by 3 and ask students to make up a word problem that would be solved using it.	
Multiplication facts	Mental Math 24
Continue to provide opportunities for students to practice the multiplication facts for 2 and 3.	
Use fact cards for the multiplication facts for 2 and 3. Have all students stand up. Hold up one card. The student who gives the correct answer first can come up and hold up the next card. Each student that gives the correct answer first replaces the student up front, who then sits down until all students get to hold up the next card. The last student gets to hold up a card for the teacher to answer.	

Enrichment	
Ask students to come up with as many true statements for 3 x 8 they can think of. Some examples are shown here. Repeat with 3 x 7.	3 x 8 = ? 3 x 8 = 24 3 x 8 = 8 x 3 3 x 8 = 8 + 8 + 8 3 x 8 = 20 + 4 3 x 8 = 30 – 3 – 3
Write the problems shown here and ask students if they are true or false.	3 x 2 = 3 + 3 ? (True) 3 x 6 = 15 + 6 ? (False) 3 x 7 = 7 x 3 ? (True) 2 x 9 = 3 x 6 ? (True) 8 x 3 = 30 – 6 ? (True) 3 x 7 = 2 x 8 ? (False)
Practice	WB Exercise 50, pp. 136-137 WB Exercise 53, p. 142 WB Exercise 54, pp. 143-144

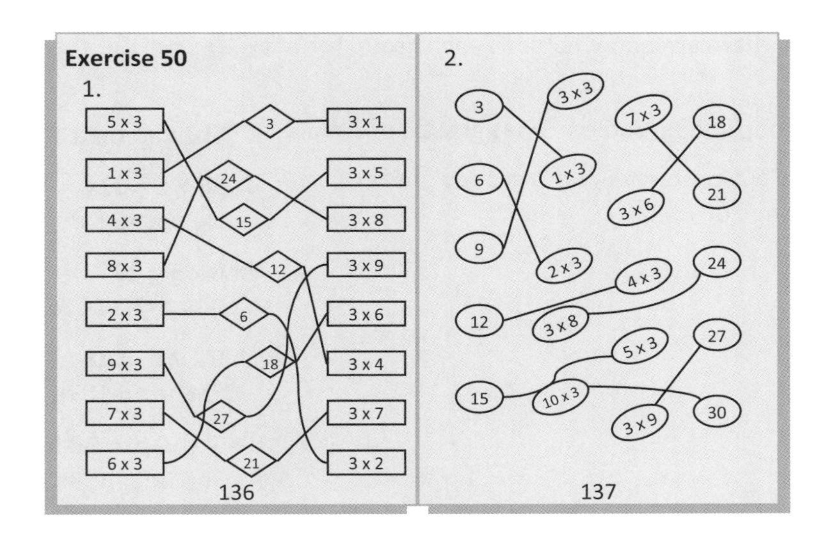

Exercise 53

1. 12 24 18

 9 21 27

 15 12 27

 30 24 18

142

Exercise 54

1. 3 x 10 = 30
 30

2. 3 x 8 m = 24 m
 24 m

3. 3 x $5 = $15
 $**15**

143

4. 7 x 3 = 21
 21

5. 9 in. x 3 = 27 in.
 27 in.

6. 3 kg x 6 = 18 kg
 18 kg

144

6.2e Practice

Objectives

- Practice multiplication facts for 2 and 3.
- Solve word problems.

Note

Students should attempt to answer multiplication problems from memory as much as possible. Continue to provide fact practice as you proceed.

Practice	**Text p. 99, Practice 6C**
Have students solve the word problems and then discuss their solutions. They should write multiplication equations for each of the word problems. If needed, they can draw simple diagrams to help them interpret the problems. Discuss how many groups there are and what is in each group for selected problems. In problem 3 there are groups of 3. Problem 9 has groups of 2 m. Problem 7 has 2 groups of $7, problem 8 has 3 groups of 6 legs, and problem 10 has 3 groups of 10 lb.	1. (a) 2 (b) 3 (c) 8 2. (a) 10 (b) 12 (c) 18 3. (a) 16 (b) 27 (c) 9 4. (a) 20 (b) 24 (c) 21 5. (a) 14 (b) 15 (c) 18 6. 3 x 5 = **15** She can read 15 books in 5 weeks. 7. $7 x 2 = **$14** He pays $14. 8. 6 x 3 = **18** 3 bees have 18 legs. 9. 9 x 2 m = **18 m** She used 18 m of lace. 10. 3 x 10 lb = **30 lb** She bought 30 lb of rice flour.
Multiplication facts	Mental Math 25-27
Continue to provide opportunities for students to practice the multiplication facts for 2 and 3.	
Practice	WB Exercise 55, pp. 145-147

Enrichment

Material: Number cards 1-30.

Procedure: Draw two overlapping ovals or circles. Explain that numbers that you get when you multiply 2 by some number go in the "x 2" circle and numbers that you get when you multiply 3 by some number go in the "x 3" circle. Numbers that could go in either circle go in the overlap. Numbers that don't go in either circle go in the space outside the circles. Have students place the cards correctly.

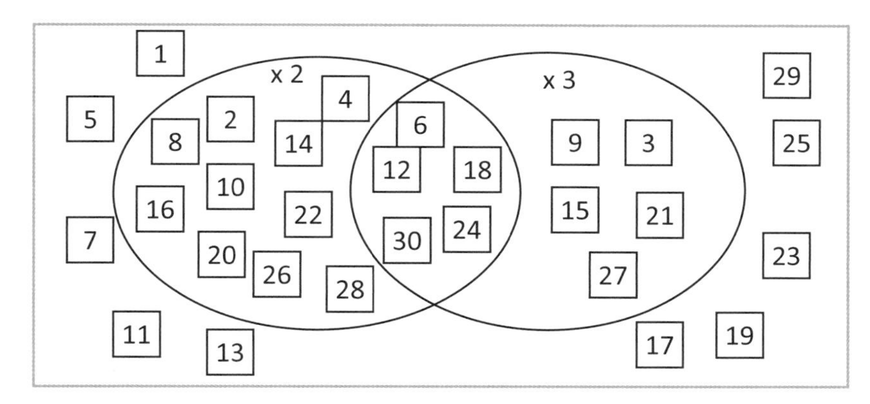

Group Game

Material: Multiplication Game Board (appendix p. a37), number cube labeled "x2" on 3 sides and "x3" on 3 sides, number cards 1-10, 4 sets per group (or 10-sided die), counters (different color for each player).

Procedure: Shuffled cards are placed face down. Players take turns drawing a card, throwing the number cube, multiplying the card by the number on the cube, and placing their counter on a square with the answer. If all answers are covered already, the player draws another card. The first player to get three in a row wins.

Exercise 55

1.

6	12
18	24
30	10
8	15
14	21
27	12

145

2. 3 x 8 = 24
24

3. 3 x 7 = 21
21

4. 2 x 7 = 14
14

146

5. 2 x 10 kg = 20 kg
20 kg

6. 8 x 2 m = 16 m
16 m

7. 3 x $6 = $18
$**18**

147

Dividing by 2 and 3

Objectives

- Use multiplication facts for 2 and 3 to divide by 2 and 3.
- Practice division facts for 2 and 3.
- Solve word problems.

Material

- Counters
- Fact cards for division by 2 and 3 with separate answer cards
- Mental Math 28-36
- Appendix pages a38-a42

Prerequisites

Since students will be finding answers to division problems using multiplication facts, they need to know the multiplication facts for 2 and 3 well.

Students need to understand the division concepts from the previous unit: that division involves either making groups and finding the number of groups, or sharing into equal groups and finding the number in each group.

Notes

In Part 3, students will relate multiplication by 2 to division by 2 and use multiplication facts to solve division by 2 problems. In Part 4, they will relate multiplication by 3 to division by 3 and use multiplication facts to solve division by 3 problems.

Even though grouping can be thought of as repeated subtraction, this curriculum does not emphasize repeated subtraction as a method for doing division. Repeated subtraction only represents one division situation, and students need to keep track of how many times they are subtracting to get the answer. Instead, this curriculum focuses on the equal parts-whole relationship between multiplication and division.

In multiplication, we know the number of parts and how much is in each part. Division is the reverse; we know the total, and the number in each part, and need to find the number of parts. Or we know the total and the number of parts, and need to find the number in each part.

To find the answer to 24 ÷ 3 we can think of *what* times 3 = 24? 8 x 3 = 24, so 24 ÷ 3 = 8.

Division is therefore not isolated from multiplication, and division facts are not a whole separate set of facts to be memorized. If multiplication facts are known, they can be applied to division facts.

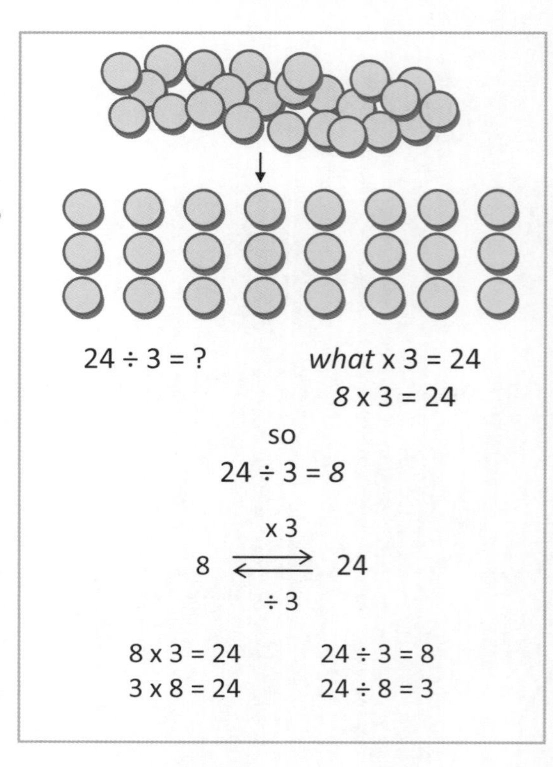

Provide students with plenty of fact practice. This can take the form of games, Sprints (answer as many facts as possible in 1 minute) or class activities. Some suggestions are offered in this guide. Feel free to be creative and add your own ideas to allow students to increase their skills in a fun way.

6.3a Divide by 2

Objectives

- Relate division by 2 to multiplication by 2.

Note

If students are competent with the multiplication facts for 2, and have a thorough understanding of the concept of multiplication and division, they will be able to answer division by 2 problems by thinking of the related multiplication problem. Less capable students can use counters or linking cubes longer.

Divide by 2	
Display counters or other objects in two groups of five. Ask students to tell you the number of counters in each group, the number of groups, and the total number of counters. Then write a multiplication equation.	 
Using an arrow to go from 5 to 10, and writing x 2 above the arrow, tell students that we multiplied 5 by 2 to get 10.	$5 \times 2 = 10$ $5 \xrightarrow{\times 2} 10$
Put the two groups of counters together and tell them that we start with 10 counters and put them in two groups. How many will be in each group? Show the two groups and write the division equation.	$10 \div 2 = 5$
Then, on the arrow diagram, draw another arrow going from the 10 to the 5 and label it with ÷ 2. Tell students we divided by 2 to get 5.	$5 \underset{\div 2}{\overset{\times 2}{\rightleftarrows}} 10$
Discuss how division is like going backwards compared to multiplication. In multiplication, we have the number of groups and the number in each group and we have to find the total. In division, we have the total and the number of groups, and have to find the number in each group (or we have the total and the number to go in each group, and have to find the number of groups).	$10 \div 2 = ?$ *what* x 2 = 10 *5* x 2 = 10 so 10 ÷ 2 = *5*
In summary, with a division problem such as 10 ÷ 2, if we remember what x 2 is 10, we will know the answer to the division problem.	
Discussion	**Text p. 100**
In the first picture we want to put 6 flowers into 2 vases. How many go in each vase? We are dividing 6 by 2 and the answer is 3. This number times 2, the number of vases, is 6. Similarly, the answer to 10 ÷ 2, the number of flowers that will be in each vase in the second picture, is the number we multiply by 2 to get 10.	3 5

Division Table	Appendix p. a38
Have students fill out a copy of the chart on appendix p. a38 in the appendix. Have students who are still struggling illustrate each pair of problems with counters.	1 x 2 = 2 2 ÷ 2 = 1 2 x 2 = 4 4 ÷ 2 = 2 3 x 2 = 6 6 ÷ 2 = 3 4 x 2 = 8 8 ÷ 2 = 4 5 x 2 = 10 10 ÷ 2 = 5 6 x 2 = 12 12 ÷ 2 = 6 7 x 2 = 14 14 ÷ 2 = 7 8 x 2 = 16 16 ÷ 2 = 8 9 x 2 = 18 18 ÷ 2 = 9 10 x 2 = 20 20 ÷ 2 = 10 11 x 2 = 22 22 ÷ 2 = 11 12 x 2 = 24 24 ÷ 2 = 12
Assessment	**Text p. 101**
	1. (a) 4 (b) 7 2. 8 10 8 10
Write some division by 2 problems on the board and ask students to try to find the answer by thinking of the related multiplication problem.	
Practice	WB Exercise 56, pp. 148-149

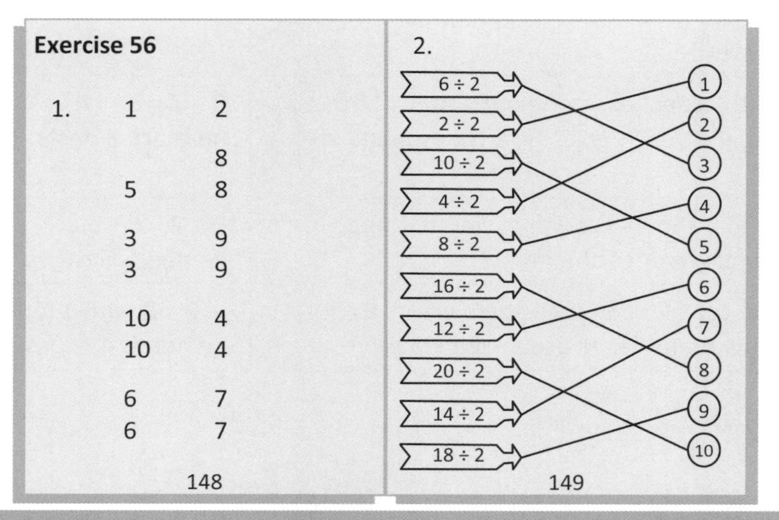

Exercise 56

1. 1 2
8
5 8
3 9
3 9
10 4
10 4
6 7
6 7

148

2.

149

6.3b Practice

Objectives

- Practice division facts for 2.
- Solve word problems.

Note

For the textbook problems, you may want to write the problems on the board rather than have students use the textbook. Capable students should be able to solve the problems easily. For less capable students, use drawings to help them come up with the equations. Encourage students to think of the related multiplication problem to find the answer. They may need to act out the problem with manipulatives.

For problems that involve measurement, encourage students to include units of measurement with the numbers.

Word Problems	**Text pp. 102-103**
Task 3: This is a sharing problem.	3. 8 ÷ 2 = **4** 4 8 ?
Task 4: This is a grouping problem. Since the number of groups is not known, a number bond drawing can indicate that we need to know the number of lines.	4. 14 ÷ 2 = **7** 7 14 2 ?
Task 5: This is a grouping problem; we need to "group" 2 m. You can provide students with a piece of string 12 meters long and let them act this problem out.	5. 12 m ÷ 2 m = **6** 6
Task 6: This is a sharing problem. You can have students act this out as well to see the difference between this problem and the previous one.	6. 18 m ÷ 2 = **9 m** 9 m
Assessment	Appendix p. a39
Use appendix p. a39 or similar problems. Ask students to write an equation and find the answer.	
1. There are 16 posts with branches in one of Harriet's bird cages. 2 birds are sitting on each post. How many posts are there?	16 ÷ 2 = 8 There are 8 posts.
2. Harriet took 10 birds to a show. She wants to put 2 birds in each cage. How many cages does she need?	10 ÷ 2 = 5 She needs 5 cages.
3. For another show, she took 7 birds. She again wants to put 2 birds in each cage. How many cages does she need?	7 ÷ 2 = 3 with 1 left over. She needs 4 cages.

Fact practice	Mental Math 28-29
Students need to continue to practice math facts. You can adapt any of the previous games. You can hold up number cards for the even numbers through 20 or 24 and have students divide the number by 2.	
Practice	WB Exercise 57, pp. 150-151

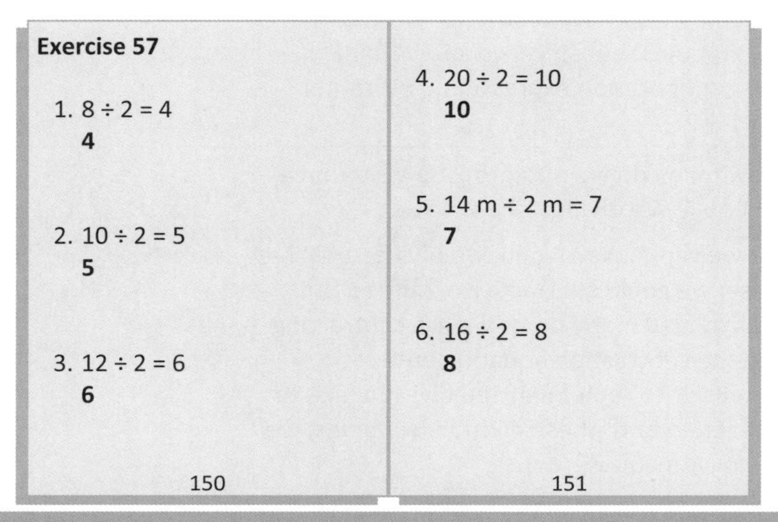

Exercise 57

1. $8 \div 2 = 4$
   4
2. $10 \div 2 = 5$
   5
3. $12 \div 2 = 6$
   6

150

4. $20 \div 2 = 10$
   10
5. $14 \text{ m} \div 2 \text{ m} = 7$
   7
6. $16 \div 2 = 8$
   8

151

6.3c Practice

Objectives

- Practice multiplication and division facts for 2.
- Solve word problems.

Note

Continue to provide practice with the multiplication facts for 2 and 3 and the division facts for 2. Use activities, such as sprints with the mental math sheets that encourage students to answer quickly, so that they move away from simply counting by twos or threes or having to think of the related multiplication fact.

Four operations	
Use counters to review what students have learned so far in solving word problems which involve the four operations.	
There are 4 red counters and 3 blue counters. How many counters are there in all? We add: 4 + 3 = 7. For addition, we are given two or more unequal parts (usually), and asked to find the total. We could also add if we were given two or more equal parts.	4 3 ? 4 + 3 = 7
There 18 counters. 12 of them are red. How many are not red? We subtract: 18 – 12 = 6. For subtraction, we are given the total, 18 counters, and a part, 12 red counters, and have to find the other part.	18 12 ? 18 – 12 = 6
There are 2 bowls, each with 8 counters. How many counters are there in all? We want to find the total. We multiply: 8 x 2 = 16. For multiplication, we have the number of groups, or equal parts, the amount in each part, and need to find the total. Point out that we could also add with this problem; 8 + 8 = 16. But if there are lots of equal groups, it is much faster to multiply if we know the multiplication facts. For example, we have 8 bowls, each with 2 counters. We can add twos 8 times, but since we know that 8 x 2 = 16, it is easier to write the multiplication expression 8 x 2 rather than 2 + 2 + 2 + 2 + 2 + 2 + 2 + 2.	8 8 ? 8 x 2 = 16
There are 14 counters. How many do we put into 2 bowls to have the same number in each bowl? We divide: 14 ÷ 2 = 7. Or, how many groups of 2 we can make? Again, we divide: 14 ÷ 2 = 7. Point out that in this case, we could subtract two 7 times, but then we have to keep track of how many times we are subtracting 2, and we can't even write the subtraction equation without already knowing that. It is easier to remember the division fact, or the related multiplication fact. In both these division examples, we are given a total and end up with equal groups.	14 ? ? 14 ÷ 2 = 7

Word problems	Appendix p. a40
Ask students to solve the problems, and then discuss them in terms of whether we are given a whole or parts, and whether the parts involve equal groups.	
1. Harriet bought 18 pounds of birdseed. The birdseed came in 2 bags. How many pounds did each bag weigh?	18 lb ÷ 2 = 9 lb Each bag weighed 9 lb.
2. Harriet used 6 pounds of the birdseed. How many pounds are left?	18 lb – 6 lb = 12 lb 12 lb are left.
3. For the remaining birdseed, she used 2 pounds a day. How many days did it take to use up the birdseed?	12 lb ÷ 2 lb = 6 It took 6 days to use up the birdseed.
4. She needed to buy more birdseed. This time she bought 18 pounds and an additional 7 pounds of birdseed. How many pounds of birdseed does she have now?	18 lb + 7 lb = 25 lb She has 25 lb now.
5. Harriet bought two new birds. One cost $398 and the other cost $276. How much did they both cost?	$398 + $276 = $674 They cost $674 altogether.
6. Harriet had $800 to spend on the birds. How much money did she have left over?	$800 – $674 = $126 She had $126 left over.
Math fact practice	Mental Math 30
Assessment	**Text p. 104, Practice 6D**
	1. (a) 8 (b) 10 (c) 4 2. (a) 4 (b) 5 (c) 2 3. (a) 12 (b) 18 (c) 16 4. (a) 6 (b) 9 (c) 8 5. (a) 7 (b) 1 (c) 10 6. 20 ÷ 2 = **10** There were 10 chairs in each row. 7. $18 ÷ $2 = **9** He took 9 days to save $18. 8. 2 x $5 = **$10** She paid $10 for the grapes. 9. 16 m ÷ 2 = **8 m** Each piece is 8 m long. 10. 14 ÷ 2 = **7** She needs 7 boxes.

6.4a Divide by 3

Objectives

- Relate division by 3 to multiplication by 3.

Note

The concepts in this lesson are not new, so it may go quickly. You can spend more time with fact memorization, and add in a few problems involving addition and subtraction of 3-digit numbers for review.

Since there are 3 feet in a yard, you can include problems involving conversion from yards to feet or feet to yards, if you think students are capable. Conversion of measurements is taught in *Primary Mathematics* 3A.

Divide by 3	
Display 24 counters or other objects. Tell students you want to divide the counters into groups of 3. Ask them how many groups you will get. Divide the counters to show that there will be 8 groups. Arrange them in an array. Point out that we now have 3 multiplied by 8. The answer to 3 x 8 is therefore the total we started with. So if we can remember *what* x 8 is 24, then we know how many groups we will end up with. We can answer division problems by remembering the related multiplication problem. You can show the relationship with arrows, as in lesson 6.3a. Ask students to write 2 multiplication and 2 division equations for the array. (Students will not be learning division facts for 8 now, but they should still be able to write the equation from the array.) Repeat with a few other examples. Write some division by 3 problems on the board and get students to think of the related multiplication problem to solve them.	24 ÷ 3 = ? *what* x 3 = 24 *8* x 3 = 24 so 24 ÷ 3 = *8* x 3 8 ⇄ 24 ÷ 3 8 x 3 = 24 24 ÷ 3 = 8 3 x 8 = 24 24 ÷ 8 = 3
Division table	Appendix p. a41
Have students fill out a copy of the chart on appendix p. a41.	1 x 3 = 3 3 ÷ 3 = 1 2 x 3 = 6 6 ÷ 3 = 2 3 x 3 = 9 9 ÷ 3 = 3 4 x 3 = 12 12 ÷ 3 = 4 5 x 3 = 15 15 ÷ 3 = 5 6 x 3 = 18 18 ÷ 3 = 6 7 x 3 = 21 21 ÷ 3 = 7 8 x 3 = 24 24 ÷ 3 = 8 9 x 3 = 27 27 ÷ 3 = 9 10 x 3 = 30 30 ÷ 3 = 10 11 x 3 = 33 33 ÷ 3 = 11 12 x 3 = 36 36 ÷ 3 = 12

Optional: Remind students that a yard is the same as 3 feet. Ask them how many yards are the same as 30 feet, or other multiples of 3.	$30 \div 3 = 10$ There are 10 yards in 30 feet.
Challenge: Remind students that a foot is the same as 12 inches. Ask students how many yards are in 36 inches. Get them to work on the problem individually for a while. Students may find that they can divide 36 by 12 by remembering that 3 x 12 = 36. This will give them the number of feet in 36 inches. Since this is 3 feet, then 36 inches is 1 yard.	12 x 3 = 36, so there are 3 feet in 36 inches, or 1 yard in 36 inches.
Assessment	**Text p. 105**
	6; 6 1. 8 5 8 5 7 9 7 9
Fact practice	Mental Math 31-32
Practice	WB Exercise 58, pp. 152-153

Group Game

Material: Copies of the Division Game Board (Appendix p. a42), fact cards for division by 2 and 3 (without answers), counters, different color for each player.

Procedure: Shuffled cards are placed face down. Players take turns drawing a card, and placing a counter on a square with the answer. If all answers are covered already, the player draws another card. The first player to get three in a row wins. The starred spaces are "wild" and can complete any player's row.

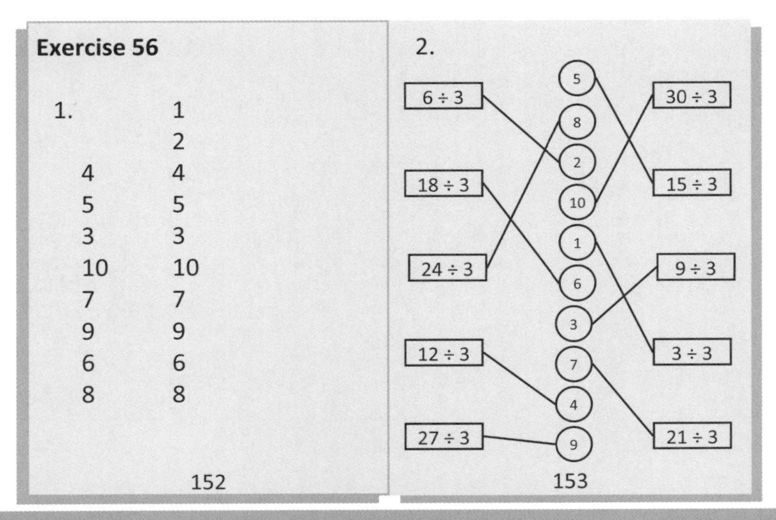

6.4b Practice

Objectives

- Practice division facts for 2 and 3.
- Solve word problems.

Note

You may want to take several days for this lesson, providing plenty of fact practice in the form of games, activities, or mental math sprints.

You may want to write the problems from p. 106 of the textbook on the board. Some students may need to draw simple pictures or use manipulatives. Encourage students to think of the related multiplication problem to find the answer, if they have not yet memorized the division facts.

Word Problems	**Text p. 106**
Task 2: This is a grouping problem. Students must think of the apples as being grouped by 3. Each group is then \$1, so the amount he pays is numerically equivalent to the number of groups. Task 3: This is a sharing problem. \$24 is shared into three equal groups, and the amount in each group is how much each child pays.	2. 30 ÷ 3 = **10** **10** x 3 = 30 \$10 3. 24 ÷ 3 = **8** **8** x 3 = 24 \$8
Write some multiplication or division expressions on the board and get students to make up some word problems to go with them.	
Assessment	**Text p. 107, Practice 6E**
	1. (a) 12 (b) 18 (c) 15 2. (a) 4 (b) 6 (c) 5 3. (a) 27 (b) 21 (c) 24 4. (a) 9 (b) 7 (c) 8 5. (a) 3 (b) 2 (c) 10 6. 30 ÷ 3 = **10** There were 10 bottles in each box. 7. \$18 ÷ 3 = **\$6** 1 kg costs \$6. 8. 15 ÷ 3 = **5** There were 5 soldiers in each row. 9. 9 x \$3 = **\$27** He paid \$27 altogether. 10. 24 ÷ 3 = **8** There are 8 beads on each string.

Math fact practice	Mental Math 33
Practice	WB Exercise 59, pp. 154-155 WB Exercise 60, pp. 156-157

Game

Material: Large fact cards and matching answer cards. Punch a hole in the answer cards and loop a string through the hole so students can wear them, or fold over masking tape to stick the card to the students' shirts so that they can display their cards but leave their hands free. They could also hold the cards in one hand and point with the other.

Procedure: Students line up in a horseshoe shape around the teacher, in the numerical order of their answer cards. The teacher shows a fact card, says the answer out loud, and points to the student with the correct answer. The teacher shows another fact card. The student first pointed to must say the answer out loud and point to the student with the correct answer to the new fact. This student will now be the pointer for the next fact card the teacher holds up. Students see how fast they can keep the game going. Once they get used to the game, they can do it silently, just pointing to the student with the answer.

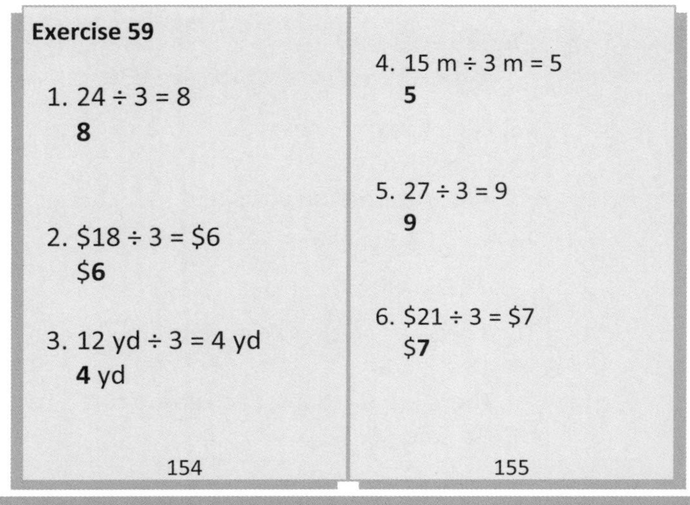

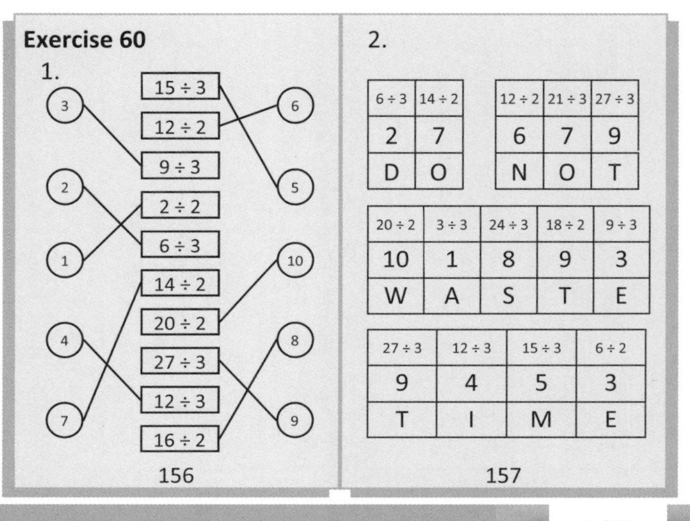

6.4c Practice

Objectives

- Practice multiplication and division facts for 2 and 3.
- Solve word problems.

Note

You may want to take several days for this lesson.

Students should attempt to answer the division problems from memory as much as possible.

Continue fact practice. You may also want to include some earlier mental math pages for addition and subtraction.

Word problem	
Write the following word problem on the board. Ask students to work on it, and then discuss their solutions. If they have trouble, include the hint. ⇒ There are 4 tanks of water, labeled A, B, C, and D. Three of them have the same capacity and one tank, D, has a capacity of 14 gallons. The total capacity of all four tanks is 41 gallons. What is the capacity of tank A? Hint: First find the capacity of tanks A, B, and C together.	A B C D 14 gal Capacity of A, B, and C: 41 gal – 14 gal = 27 gal Capacity of A: 27 gal ÷ 3 = 9 gal The capacity of tank A is 9 gallons.
Assessment	**Text p. 108, Practice 6F**
Students should write equations for each of the word problems. You can write the problems on the board. If needed, they can draw simple diagrams to help them interpret the problems.	1. (a) 5 (b) 7 (c) 4 2. (a) 3 (b) 5 (c) 4 3. (a) 6 (b) 8 (c) 10 4. (a) 6 (b) 8 (c) 7 5. (a) 9 (b) 10 (c) 9 6. 24 cm ÷ 3 = **8 cm** Each piece is 8 cm long. 7. \$30 ÷ \$3 = **10** It took him 10 weeks to save \$30. 8. Each kilogram costs \$2. \$2 x 7 = **\$14** The watermelon costs \$14. 9. 16 kg ÷ 2 kg = **8** He got 8 bags. 10. 3 pears cost \$1, so we need to group by 3. 18 ÷ 3 = **6** There are 6 groups. Each group cost \$1. He paid \$6.

Math fact practice	Mental Math 34-36
Enrichment	
Write the problems on the right on the board, and ask students to fill in the blanks.	3 + 3 + 3 + 3 + 3 = ______ x 3 (5) 18 + 6 = ______ x 3 (8) 9 x 2 = 9 + ______ (9) 9 x 3 = 9 + ______ (18) 30 ÷ 3 = _____ ÷ 2 (20)
Write the problems on the right on the board, and ask students to fill in the circles with >, <, or =. Students who understand the concepts will not need to evaluate both sides to determine what symbol should be used. For example, they should not have to find what 9 x 9 or 8 x 9 are.	2 x 7 ◯ 3 x 7 (<) 3 + 3 + 3 + 3 ◯ 2 + 2 + 2 + 2 + 2 + 2 (=) 20 – 6 ◯ 2 x 9 (<) 387 – 88 ◯ 300 (<) 9 x 9 ◯ 8 x 9 (>) 6 tens ÷ 2 ◯ 3 tens (=) 401 ◯ 302 + 98 (>)
Practice	WB Exercise 61, pp. 158-159 WB Exercise 62, pp. 160-162

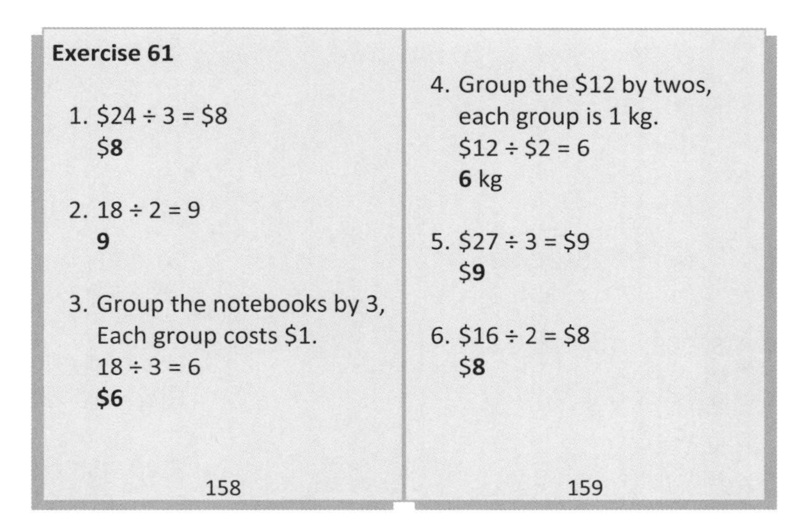

Exercise 61

1. $24 ÷ 3 = $8
   $8

2. 18 ÷ 2 = 9
   9

3. Group the notebooks by 3,
   Each group costs $1.
   18 ÷ 3 = 6
   $6

158

4. Group the $12 by twos,
   each group is 1 kg.
   $12 ÷ $2 = 6
   6 kg

5. $27 ÷ 3 = $9
   $9

6. $16 ÷ 2 = $8
   $8

159

Exercise 62

1. 10 18 4 8

   12 7 10 14

   16 7 9 27

160

2. 15 cm ÷ 3 = 5 cm
   5 cm

3. 3 x 7 = 21
   21

4. 3 x 4 = 12
   12

161

3. 6 x 3 yd = 18 yd
   18 yd

4. $24 ÷ 3 = $8
   $8

5. $18 ÷ 2 = $9
   $9

162

6.4d Review

Objectives

- Review all previous topics.

Note

This review will take several days. You can do half of a review each day, and assign 3 pages from the workbook for homework for assessment.

If certain topics need to be re-taught or reviewed more extensively, you can adapt the lessons pertaining to those topics, depending on what the misunderstandings are.

Review	Text pp. 109-110, Review B	Text pp. 111-112, Review C
	1. (a) 606 (b) 855 (c) 440 2. (a) Two hundred fifty (b) Seven hundred forty-four (c) Three hundred seven (d) Nine hundred twenty-two 3. (a) 213 (b) 449 (c) 799 (d) 325 4. (a) 15, 18, 21; 27 (b) 470, 460, 450; 430 5. 909, 912, 928, 930 6. (a) 8 (b) 8 (c) 90 (d) 9 (e) 70 (f) 5 (g) 100 (h) 50 7. (a) 4 kg (b) 350 g 8. brush 1 cm 9. 5 x 3 = **15** She bought 15 apples. 10. 128 + 25 = **153** There are 153 girls. 11. 142 cm – 14 cm = **128 cm** Her brother is 128 cm tall. 12. 24 ÷ 3 = **8** She used 8 plates.	1. (a) 408 (b) 590 (c) 555 2. (a) 78 (b) 703 (c) 734 3. (a) 18 (b) 24 (c) 16 4. (a) 9 (b) 8 (c) 8 5. (a) 7 (b) 10 (c) 10 6. (a) 689 (b) 505 (c) 40 (d) 0 7. (a) < (b) > (c) > (d) < 8. (a) 130 g (b) 120 (250 g – 130 g = 120 g) 9. 128 + 94 + 46 = **268** There are 268 members. 10. \$120 – \$89 = **\$31** She needs \$31 more. 11. 3 x 6 = **18** She bought 18 cakes. 12. \$27 ÷ 3 = **\$9** Each child received \$9. 13. 820 + 95 = **915** There are 915 coconut trees. 14. \$145 + \$65 = **\$210** The camera costs \$210. 15. 18 m ÷ 2 m = **9** He made 9 shirts.

Assessment	WB Review 6, pp. 163-168 WB Review 7, pp. 169-174

Review 6

1. (a) eight hundred fifty-seven
   (b) six hundred forty-four

2. 6 x 3 = 18 18 ÷ 6 = 3
   3 x 6 = 18 18 ÷ 3 = 6

3. (a) 456 – 50 = 406
   or 456 – 406 = 50
   (b) 275 + 325 = 600
   or 325 + 275 = 600
   (c) 3 x 9 = 27 or 9 x 3 = 27
   (d) 18 ÷ 2 = 9 or 18 ÷ 9 = 2

163

4.

950	960	970	**980**	990	**1000**
	959			989	
	958			**988**	
	957			987	
946	**956**	966	976	**986**	**996**

5. (a) 7; 7
   (b) 6; 6

6. 829, 831, 846, 852

7. (a) 7
   (b) 50

164

8.

440	96	230
103	450	636
330	352	840
320	455	518

Head of a lion

165

9.

8 ↓ 10 ↓ 12 ↓ 6 ↓ 4 ↓ 8 ⟶ 16 ↑ 14 ↑ 7 ↑ 9 ↑ 18 ↑ 20 ⟶ 10 ↓ 8 ↓ 6 ↓ 12 ↓ 10 ↓ 20

166

10. 402 – 35 = 367
    367

11. 6 x \$4 = \$24
    \$24

12. 24 ÷ 3 = 8
    8

167

13. 460 – 295 = 165
    165

14. 18 ÷ 2 = 9
    9

15. 285 + 168 = 453
    453

168

Review 7

1. (a) 550
   (b) 929
2. (a) seven hundred forty-four
   (b) eight hundred six
3. (a) < (b) <
   (c) > (d) >
   (e) > (f) <
4. (a) 879 (b) 760
   (c) 504 (d) 90
   (e) 607 (f) 10

169

5. (a) 3 cm
   (b) 12 cm

6. (a) 4 kg
   (b) \$12

7. (a) lb
   (b) in.
   (c) oz
   (d) yd
   (e) ft

170

8.
   (a) 473 (b) 516
   (c) 852 (d) 800
   (e) 340 (f) 126
   (g) 358 (h) 575

171

9.
   (a) 6 (b) 15
   (c) 21 (d) 18
   (e) 9 (f) 10
   (g) 16 (h) 27
   (i) 1 (j) 2
   (k) 5 (l) 9
   (m) 1 (n) 5
   (o) 8 (p) 7

172

10. 12 lb ÷ 2 = 6 lb
    6 lb

11. 6 x \$3 = \$18
    \$18

12. 18 yd – 3 yd = 15 yd
    15 yd

173

13. \$500 – \$264 = \$236
    \$236

14. 85 + 46 = 131
    131

15. 420 - 285 = 135
    135

174

Answers to Mental Math

Mental Math 1	Mental Math 2	Mental Math 3		Mental Math 4	Mental Math 5	Mental Math 6
25 + 2 = **27**	5 + 3 = **8**	10 – 7 = **3**		8 – 2 = **6**	9 + 5 = 10 + **4**	9 + 6 = **15**
44 + 1 = **45**	4 + 1 = **5**	6 – 2 = **4**		7 – 6 = **1**	7 + 4 = 10 + **1**	8 + 9 = **17**
69 + 3 = **72**	2 + 6 = **8**	8 – 1 = **7**		9 – 2 = **7**	8 + 3 = 10 + **1**	5 + 8 = **13**
20 – 1 = **19**	4 + 3 = **7**	10 – 3 = **7**		7 – 2 = **5**	6 + 7 = 10 + **3**	6 + 6 = **12**
72 – 3 = **69**	2 + 8 = **10**	3 – 2 = **1**		9 – 8 = **1**	6 + 7 = 3 + **10**	9 + 2 = **11**
55 – 2 = **53**	5 + 5 = **10**	5 – 3 = **2**		10 – 1 = **9**	6 + 8 = **4** + 10	4 + 7 = **11**
36 + 3 = **39**	3 + 7 = **10**	10 – 8 = **2**		7 – 5 = **2**	8 + 9 = 10 + **7**	8 + 6 = **14**
29 + 2 = **31**	2 + 2 = **4**	6 – 3 = **3**		5 – 2 = **3**	3 + 8 = 10 + **1**	7 + 5 = **12**
61 – 2 = **59**	1 + 8 = **9**	9 – 1 = **8**		10 – 9 = **1**	8 + **4** = 10 + 2	3 + 9 = **12**
34 + 3 = **37**	5 + 4 = **9**	5 – 5 = **0**		9 – 7 = **2**	7 + **9** = 10 + 6	6 + 7 = **13**
27 – 3 = **24**	6 + 3 = **9**	8 – 4 = **4**		5 – 4 = **1**	12 – 6 = 4 + **2**	5 + 9 = **14**
70 – 2 = **68**	7 + 2 = **9**	10 – 6 = **4**		8 – 3 = **5**	13 – 7 = 3 + **3**	7 + 8 = **15**
28 + 3 = **31**	2 + 5 = **7**	6 – 4 = **2**		10 – 2 = **8**	16 – 9 = 1 + **6**	8 + 4 = **12**
39 – 2 = **37**	3 + 2 = **5**	9 – 3 = **6**		8 – 6 = **2**	12 – 7 = **3** + 2	9 + 4 = **13**
100 – 1 = **99**	9 + 1 = **10**	8 – 7 = **1**		9 – 4 = **5**	14 – 9 = **1** + 4	7 + 7 = **14**
98 + 2 = **100**	2 + 4 = **6**	9 – 6 = **3**		7 – 3 = **4**	17 – 8 = **2** + 7	6 + 5 = **11**
35 + 1 = **36**	3 + 3 = **6**	7 – 4 = **3**		10 – 4 = **6**	11 – 6 = **4** + 1	9 + 7 = **16**
21 – 2 = **19**	1 + 5 = **6**	10 – 5 = **5**		6 – 5 = **1**	15 – 7 = **3** + 5	7 + 2 = **9**
44 + 2 = **46**	4 + 4 = **8**	8 – 5 = **3**		9 – 5 = **4**	17 – 6 = 10 + **1**	8 + 8 = **16**
61 – 3 = **58**	6 + 4 = **10**	4 – 2 = **2**		4 – 3 = **1**	18 – 5 = **10** + 3	3 + 8 = **11**

Mental Math 7	Mental Math 8	Mental Math 9		Mental Math 10	Mental Math 11	Mental Math 12
17 – 8 = **9**	11 – 7 = **4**	13 + 3 = **16**		563 + 3 = **566**	30 + 60 = **90**	913 + 16 = **929**
11 – 9 = **2**	14 – 6 = **8**	12 – 2 = **10**		642 – 2 = **640**	43 + 20 = **63**	626 + 150 = **776**
14 – 5 = **9**	12 – 3 = **9**	9 + 6 = **15**		801 + 20 = **821**	77 + 2 = **79**	461 + 138 = **599**
15 – 6 = **9**	11 – 4 = **7**	14 – 3 = **11**		349 – 30 = **319**	61 + 24 = **85**	176 + 202 = **378**
12 – 8 = **4**	14 – 9 = **5**	10 + 3 = **13**		430 + 300 = **730**	80 – 30 = **50**	602 + 326 = **928**
12 – 4 = **8**	18 – 4 = **14**	8 – 4 = **4**		817 – 200 = **617**	73 – 60 = **13**	140 + 705 = **845**
15 – 7 = **8**	12 – 5 = **7**	9 + 6 = **15**		209 + 2 = **211**	58 – 4 = **54**	483 + 201 = **684**
12 – 9 = **3**	13 – 6 = **7**	13 – 5 = **8**		872 – 300 = **572**	95 – 61 = **34**	170 + 616 = **786**
11 – 8 = **3**	12 – 3 = **9**	15 – 8 = **7**		459 – 30 = **429**	23 + 72 = **95**	202 + 410 = **612**
11 – 3 = **8**	11 – 5 = **6**	9 + 7 = **16**		391 + 10 = **401**	85 – 4 = **81**	824 + 73 = **897**
14 – 7 = **7**	16 – 4 = **12**	8 – 3 = **5**		981 – 30 = **951**	79 – 32 = **47**	719 + 230 = **949**
16 – 9 = **7**	19 – 7 = **12**	12 – 4 = **8**		620 – 30 = **590**	21 + 63 = **84**	783 + 12 = **795**
13 – 5 = **8**	15 – 9 = **6**	15 + 2 = **17**		459 + 200 = **659**	69 – 9 = **60**	43 + 544 = **587**
16 – 7 = **9**	12 – 6 = **6**	9 + 3 = **12**		59 + 300 = **359**	43 + 34 = **77**	202 + 177 = **379**
13 – 4 = **9**	16 – 8 = **8**	8 + 2 = **10**		394 + 20 = **414**	95 – 51 = **44**	24 + 635 = **659**
13 – 8 = **5**	15 – 3 = **12**	19 – 2 = **17**		290 – 200 = **90**	48 – 40 = **8**	410 + 544 = **954**
13 – 7 = **6**	17 – 9 = **8**	8 – 2 = **6**		618 – 20 = **598**	6 + 32 = **38**	431 + 30 = **461**
14 – 8 = **6**	15 – 8 = **7**	17 – 9 = **8**		117 – 30 = **87**	80 + 14 = **94**	203 + 280 = **483**
13 – 9 = **4**	11 – 6 = **5**	10 + 7 = **17**		202 + 20 = **222**	99 – 54 = **45**	315 + 50 = **365**
12 – 7 = **5**	18 – 9 = **9**	20 – 10 = **10**		1000 – 10 = **990**	9 + 40 = **49**	300 + 539 = **839**

Answers to Mental Math

Mental Math 13	Mental Math 14	Mental Math 15	Mental Math 16	Mental Math 17	Mental Math 18
349 – 11 = **338**	10 – 3 = **7**	9 + 7 = **16**	14 – 5 = **9**	542 – 9 = **533**	461 + 300 = **761**
869 – 150 = **719**	4 + 5 = **9**	90 + 70 = **160**	94 – 5 = **89**	674 – 7 = **667**	394 – 50 = **344**
484 – 62 = **422**	6 + 7 = **13**	39 + 7 = **46**	394 – 5 = **389**	52 – 6 = **46**	490 + 2 = **492**
689 – 602 = **87**	15 – 8 = **7**	390 + 70 = **460**	940 – 50 = **890**	450 – 60 = **390**	978 – 100 = **878**
578 – 63 = **515**	34 + 23 = **57**	139 + 7 = **146**	16 – 9 = **7**	78 – 2 = **76**	342 – 4 = **338**
679 – 405 = **274**	85 – 3 = **82**	391 + 70 = **461**	36 – 9 = **27**	300 – 90 = **210**	5 + 378 = **383**
658 – 58 = **600**	400 + 200 = **600**	139 + 70 = **209**	236 – 9 = **227**	410 – 50 = **360**	200 + 459 = **659**
927 – 715 = **212**	900 – 300 = **600**	60 + 40 = **100**	360 – 90 = **270**	72 – 8 = **64**	821 – 9 = **812**
866 – 530 = **336**	400 + 50 = **450**	60 + 50 = **110**	100 – 60 = **40**	350 – 80 = **270**	891 – 20 = **871**
748 – 332 = **416**	999 – 20 = **979**	60 + 140 = **200**	700 – 60 = **640**	236 – 7 = **229**	30 + 29 = **59**
559 – 103 = **456**	459 – 8 = **451**	60 + 150 = **210**	43 – 6 = **37**	891 – 8 = **883**	8 + 64 = **72**
853 – 723 = **130**	201 + 4 = **205**	60 + 152 = **212**	430 – 60 = **370**	960 – 20 = **940**	110 – 9 = **101**
897 – 87 = **810**	564 – 40 = **524**	348 + 2 = **350**	430 – 6 = **424**	502 – 50 = **452**	110 – 90 = **20**
537 – 301 = **236**	923 + 70 = **993**	348 + 20 = **368**	40 – 8 = **32**	64 – 8 = **56**	800 + 200 = **1000**
692 – 190 = **502**	432 + 34 = **466**	570 + 30 = **600**	43 – 8 = **35**	456 – 9 = **447**	705 – 400 = **305**
295 – 244 = **51**	863 – 42 = **821**	580 + 30 = **610**	343 – 8 = **335**	730 – 50 = **680**	6 + 378 = **384**
816 – 402 = **414**	42 + 127 = **169**	580 + 34 = **614**	400 – 80 = **320**	75 – 9 = **66**	60 + 378 = **438**
749 – 46 = **703**	346 + 643 = **989**	582 + 34 = **616**	430 – 80 = **350**	97 – 1 = **96**	600 + 378 = **978**
703 – 201 = **502**	569 – 237 = **332**	269 + 2 = **271**	438 – 80 = **358**	370 – 80 = **290**	745 – 40 = **705**
937 – 24 = **913**	7 + 462 = **469**	269 + 20 = **289**	438 – 8 = **430**	821 – 50 = **771**	82 + 816 = **898**

Mental Math 19	Mental Math 20	Mental Math 21	Mental Math 22	Mental Math 23	Mental Math 24
2 x 4 = **8**	2 x 4 = **8**	2 x 3 = **6**	3 x 7 = **21**	3 x 3 = **9**	3 x 4 = **12**
2 x 3 = **6**	10 x 2 = **20**	6 x 2 = **12**	3 x 5 = **15**	3 x 2 = **6**	8 x 3 = **24**
2 x 8 = **16**	8 x 2 = **16**	1 x 2 = **2**	3 x 10 = **30**	6 x 3 = **18**	5 x 3 = **15**
2 x 10 = **20**	2 x 1 = **2**	2 x 6 = **12**	3 x 8 = **24**	1 x 3 = **3**	3 x 1 = **3**
2 x 1 = **2**	6 x 2 = **12**	5 x 2 = **10**	3 x 5 = **15**	3 x 6 = **18**	2 x 3 = **6**
2 x 9 = **18**	2 x 7 = **14**	2 x 1 = **2**	3 x 9 = **27**	5 x 3 = **15**	9 x 3 = **27**
2 x 2 = **4**	4 x 2 = **8**	2 x 2 = **4**	3 x 6 = **18**	3 x 1 = **3**	3 x 6 = **18**
2 x 6 = **12**	2 x 6 = **12**	8 x 2 = **16**	3 x 7 = **21**	2 x 3 = **6**	7 x 3 = **21**
2 x 7 = **14**	2 x 9 = **18**	2 x 7 = **14**	3 x 3 = **9**	8 x 3 = **24**	10 x 3 = **30**
2 x 5 = **10**	9 x 2 = **18**	2 x 3 = **6**	3 x 4 = **12**	3 x 7 = **21**	3 x 9 = **27**
2 x 10 = **20**	7 x 2 = **14**	2 x 4 = **8**	3 x 3 = **9**	3 x 3 = **9**	3 x 10 = **30**
2 x 8 = **16**	2 x 2 = **4**	2 x 10 = **20**	3 x 8 = **24**	3 x 4 = **12**	3 x 3 = **9**
2 x 5 = **10**	3 x 2 = **6**	7 x 2 = **14**	3 x 10 = **30**	3 x 10 = **30**	4 x 3 = **12**
2 x 9 = **18**	2 x 5 = **10**	2 x 12 = **24**	3 x 1 = **3**	7 x 3 = **21**	11 x 3 = **33**
2 x 6 = **12**	2 x 10 = **20**	4 x 2 = **8**	3 x 9 = **27**	3 x 9 = **27**	3 x 8 = **24**
2 x 7 = **14**	5 x 2 = **10**	10 x 2 = **20**	3 x 2 = **6**	10 x 3 = **30**	3 x 5 = **15**
2 x 3 = **6**	2 x 3 = **6**	8 x 2 = **16**	3 x 6 = **18**	8 x 3 = **24**	3 x 7 = **21**
2 x 4 = **8**	2 x 2 = **4**	2 x 11 = **22**	3 x 4 = **12**	3 x 8 = **24**	12 x 3 = **36**
2 x 12 = **24**	2 x 8 = **16**	9 x 2 = **18**	3 x 11 = **33**	4 x 3 = **12**	3 x 2 = **6**
2 x 11 = **22**	1 x 2 = **2**	2 x 5 = **10**	3 x 12 = **36**	3 x 5 = **15**	6 x 3 = **18**

Answers to Mental Math

Mental Math 25	Mental Math 26	Mental Math 27	Mental Math 28	Mental Math 29	Mental Math 30
2 x 6 = **12**	3 x 4 = **12**	2 x 8 = **16**	10 ÷ 2 = **5**	8 ÷ 2 = **4**	14 ÷ 2 = **7**
9 x 2 = **18**	2 x 1 = **2**	12 – 3 = **9**	14 ÷ 2 = **7**	8 ÷ 1 = **8**	2 x 6 = **12**
3 x 3 = **9**	10 x 3 = **30**	9 x 2 = **18**	8 ÷ 2 = **4**	18 ÷ 2 = **9**	16 ÷ 2 = **8**
2 x 3 = **6**	2 x 9 = **18**	2 + 10 = **12**	18 ÷ 2 = **9**	10 ÷ 1 = **10**	7 x 2 = **14**
2 x 8 = **16**	1 x 3 = **3**	3 x 11 = **33**	6 ÷ 2 = **3**	20 ÷ 2 = **10**	18 ÷ 2 = **9**
3 x 5 = **15**	5 x 2 = **10**	11 – 2 = **9**	2 ÷ 2 = **1**	4 ÷ 1 = **4**	6 ÷ 2 = **3**
2 x 7 = **14**	3 x 7 = **21**	8 x 3 = **24**	20 ÷ 2 = **10**	16 ÷ 2 = **8**	2 x 12 = **24**
9 x 3 = **27**	6 x 2 = **12**	7 + 3 = **10**	4 ÷ 2 = **2**	11 ÷ 1 = **11**	10 x 2 = **20**
10 x 2 = **20**	8 x 3 = **24**	2 + 6 = **8**	16 ÷ 2 = **8**	1 ÷ 1 = **1**	20 ÷ 2 = **10**
3 x 8 = **24**	2 x 2 = **4**	3 + 9 = **12**	12 ÷ 2 = **6**	12 ÷ 2 = **6**	8 ÷ 2 = **4**
2 x 2 = **4**	3 x 9 = **27**	1 + 3 = **4**	10 ÷ 2 = **5**	22 ÷ 2 = **11**	8 x 2 = **16**
7 x 3 = **21**	2 x 10 = **20**	9 + 2 = **11**	16 ÷ 2 = **8**	6 ÷ 1 = **6**	2 x 2 = **4**
2 x 5 = **10**	8 x 2 = **16**	8 – 3 = **5**	2 ÷ 2 = **1**	14 ÷ 2 = **7**	12 ÷ 2 = **6**
6 x 3 = **18**	3 x 2 = **6**	12 x 3 = **36**	18 ÷ 2 = **9**	24 ÷ 2 = **12**	10 ÷ 2 = **5**
4 x 2 = **8**	3 x 3 = **9**	8 – 2 = **6**	4 ÷ 2 = **2**	10 ÷ 2 = **5**	11 x 2 = **22**
3 x 10 = **30**	2 x 3 = **6**	2 x 11 = **22**	6 ÷ 2 = **3**	16 ÷ 1 = **16**	4 ÷ 2 = **2**
3 x 2 = **6**	3 x 6 = **18**	3 + 8 = **11**	20 ÷ 2 = **10**	4 ÷ 2 = **2**	4 x 2 = **8**
4 x 3 = **12**	2 x 4 = **8**	3 x 9 = **27**	14 ÷ 2 = **7**	40 ÷ 2 = **20**	2 ÷ 2 = **1**
1 x 2 = **2**	5 x 3 = **15**	12 x 2 = **24**	12 ÷ 2 = **6**	6 ÷ 2 = **3**	2 x 1 = **2**
3 x 1 = **3**	7 x 2 = **14**	10 + 3 = **13**	8 ÷ 2 = **4**	60 ÷ 2 = **30**	2 x 9 = **18**

Mental Math 31	Mental Math 32	Mental Math 33	Mental Math 34	Mental Math 35	Mental Math 36
15 ÷ 3 = **5**	18 ÷ 3 = **6**	12 ÷ 3 = **4**	12 ÷ 2 = **6**	3 – 3 = **0**	9 + 7 = **16**
24 ÷ 3 = **8**	18 ÷ 2 = **9**	9 x 2 = **18**	9 x 2 = **18**	27 ÷ 3 = **9**	13 – 6 = **7**
9 ÷ 3 = **3**	18 ÷ 1 = **18**	10 ÷ 2 = **5**	15 ÷ 3 = **5**	3 + 3 = **6**	123 + 321 = **444**
27 ÷ 3 = **9**	27 ÷ 3 = **9**	6 x 3 = **18**	8 ÷ 2 = **4**	18 + 2 = **20**	8 x 3 = **24**
6 ÷ 3 = **2**	9 ÷ 3 = **3**	12 ÷ 2 = **6**	6 x 3 = **18**	12 ÷ 2 = **6**	38 + 5 = **43**
18 ÷ 3 = **6**	12 ÷ 2 = **6**	7 x 3 = **21**	9 ÷ 3 = **3**	27 – 3 = **24**	2 x 9 = **18**
3 ÷ 3 = **1**	2 ÷ 2 = **1**	27 ÷ 3 = **9**	2 x 7 = **14**	18 – 2 = **16**	81 – 7 = **74**
21 ÷ 3 = **7**	24 ÷ 3 = **8**	12 ÷ 3 = **4**	18 ÷ 2 = **9**	12 – 3 = **9**	348 – 5 = **343**
18 ÷ 3 = **6**	36 ÷ 3 = **12**	12 x 2 = **24**	2 ÷ 2 = **1**	27 + 3 = **30**	235 + 7 = **242**
9 ÷ 3 = **3**	1 ÷ 1 = **1**	9 ÷ 3 = **3**	30 ÷ 3 = **10**	30 – 3 = **27**	999 – 123 = **876**
30 ÷ 3 = **10**	21 ÷ 3 = **7**	8 x 3 = **24**	3 x 8 = **24**	12 + 3 = **15**	3 x 5 = **15**
12 ÷ 3 = **4**	22 ÷ 2 = **11**	2 ÷ 2 = **1**	14 ÷ 2 = **7**	18 ÷ 2 = **9**	27 ÷ 3 = **9**
27 ÷ 3 = **9**	4 ÷ 2 = **2**	10 x 2 = **20**	3 x 4 = **12**	3 ÷ 3 = **1**	40 + 80 = **120**
15 ÷ 3 = **5**	12 ÷ 3 = **4**	16 ÷ 2 = **8**	5 x 2 = **10**	100 ÷ 100 = **1**	7 x 2 = **14**
21 ÷ 3 = **7**	33 ÷ 3 = **11**	12 x 3 = **36**	4 ÷ 2 = **2**	3 x 1 = **3**	120 – 90 = **30**
24 ÷ 3 = **8**	12 ÷ 1 = **12**	18 ÷ 3 = **6**	24 ÷ 3 = **8**	100 x 1 = **100**	18 ÷ 3 = **6**
6 ÷ 3 = **2**	15 ÷ 3 = **5**	7 x 2 = **14**	2 x 8 = **16**	3 ÷ 1 = **3**	683 + 40 = **723**
30 ÷ 3 = **10**	24 ÷ 2 = **12**	6 ÷ 3 = **2**	3 ÷ 3 = **1**	25 ÷ 1 = **25**	18 ÷ 2 = **9**
3 ÷ 3 = **1**	3 ÷ 3 = **1**	10 x 3 = **30**	3 x 9 = **27**	100 ÷ 1 = **100**	683 – 40 = **643**
12 ÷ 3 = **4**	14 ÷ 2 = **7**	9 x 2 = **18**	21 ÷ 3 = **7**	100 – 1 = **99**	99 + 9 = **108**

Appendix

Mental Math 1	Mental Math 2	Mental Math 3
25 + 2 = ______	5 + 3 = ______	10 – 7 = ______
44 + 1 = ______	4 + 1 = ______	6 – 2 = ______
69 + 3 = ______	2 + 6 = ______	8 – 1 = ______
20 – 1 = ______	4 + 3 = ______	10 – 3 = ______
72 – 3 = ______	2 + 8 = ______	3 – 2 = ______
55 – 2 = ______	5 + 5 = ______	5 – 3 = ______
36 + 3 = ______	3 + 7 = ______	10 – 8 = ______
29 + 2 = ______	2 + 2 = ______	6 – 3 = ______
61 – 2 = ______	1 + 8 = ______	9 – 1 = ______
34 + 3 = ______	5 + 4 = ______	5 – 5 = ______
27 – 3 = ______	6 + 3 = ______	8 – 4 = ______
70 – 2 = ______	7 + 2 = ______	10 – 6 = ______
28 + 3 = ______	2 + 5 = ______	6 – 4 = ______
39 – 2 = ______	3 + 2 = ______	9 – 3 = ______
100 – 1 = ______	9 + 1 = ______	8 – 7 = ______
98 + 2 = ______	2 + 4 = ______	9 – 6 = ______
35 + 1 = ______	3 + 3 = ______	7 – 4 = ______
21 – 2 = ______	1 + 5 = ______	10 – 5 = ______
44 + 2 = ______	4 + 4 = ______	8 – 5 = ______
61 – 3 = ______	6 + 4 = ______	4 – 2 = ______

Mental Math 4	Mental Math 5	Mental Math 6
8 – 2 = ______	9 + 5 = 10 + ______	9 + 6 = ______
7 – 6 = ______	7 + 4 = 10 + ______	8 + 9 = ______
9 – 2 = ______	8 + 3 = 10 + ______	5 + 8 = ______
7 – 2 = ______	6 + 7 = 10 + ______	6 + 6 = ______
9 – 8 = ______	6 + 7 = 3 + ______	9 + 2 = ______
10 – 1 = ______	6 + 8 = ______ + 10	4 + 7 = ______
7 – 5 = ______	8 + 9 = 10 + ______	8 + 6 = ______
5 – 2 = ______	3 + 8 = 10 + ______	7 + 5 = ______
10 – 9 = ______	8 + ______ = 10 + 2	3 + 9 = ______
9 – 7 = ______	7 + ______ = 10 + 6	6 + 7 = ______
5 – 4 = ______	12 – 6 = 4 + ______	5 + 9 = ______
8 – 3 = ______	13 – 7 = 3 + ______	7 + 8 = ______
10 – 2 = ______	16 – 9 = 1 + ______	8 + 4 = ______
8 – 6 = ______	12 – 7 = ______ + 2	9 + 4 = ______
9 – 4 = ______	14 – 9 = ______ + 4	7 + 7 = ______
7 – 3 = ______	17 – 8 = ______ + 7	6 + 5 = ______
10 – 4 = ______	11 – 6 = ______ + 1	9 + 7 = ______
6 – 5 = ______	15 – 7 = ______ + 5	7 + 2 = ______
9 – 5 = ______	17 – 6 = 10 + ______	8 + 8 = ______
4 – 3 = ______	18 – 5 = ______ + 3	3 + 8 = ______

Mental Math 7	Mental Math 8	Mental Math 9
17 – 8 = ______	11 – 7 = ______	13 + 3 = ______
11 – 9 = ______	14 – 6 = ______	12 – 2 = ______
14 – 5 = ______	12 – 3 = ______	9 + 6 = ______
15 – 6 = ______	11 – 4 = ______	14 – 3 = ______
12 – 8 = ______	14 – 9 = ______	10 + 3 = ______
12 – 4 = ______	18 – 4 = ______	8 – 4 = ______
15 – 7 = ______	12 – 5 = ______	9 + 6 = ______
12 – 9 = ______	13 – 6 = ______	13 – 5 = ______
11 – 8 = ______	12 – 3 = ______	15 – 8 = ______
11 – 3 = ______	11 – 5 = ______	9 + 7 = ______
14 – 7 = ______	16 – 4 = ______	8 – 3 = ______
16 – 9 = ______	19 – 7 = ______	12 – 4 = ______
13 – 5 = ______	15 – 9 = ______	15 + 2 = ______
16 – 7 = ______	12 – 6 = ______	9 + 3 = ______
13 – 4 = ______	16 – 8 = ______	8 + 2 = ______
13 – 8 = ______	15 – 3 = ______	19 – 2 = ______
13 – 7 = ______	17 – 9 = ______	8 – 2 = ______
14 – 8 = ______	15 – 8 = ______	17 – 9 = ______
13 – 9 = ______	11 – 6 = ______	10 + 7 = ______
12 – 7 = ______	18 – 9 = ______	20 – 10 = ______

Mental Math 10	Mental Math 11	Mental Math 12
563 + 3 = ______	30 + 60 = ______	913 + 16 = ______
642 – 2 = ______	43 + 20 = ______	626 + 150 = ______
801 + 20 = ______	77 + 2 = ______	461 + 138 = ______
349 – 30 = ______	61 + 24 = ______	176 + 202 = ______
430 + 300 = ______	80 – 30 = ______	602 + 326 = ______
817 – 200 = ______	73 – 60 = ______	140 + 705 = ______
209 + 2 = ______	58 – 4 = ______	483 + 201 = ______
872 – 300 = ______	95 – 61 = ______	170 + 616 = ______
459 – 30 = ______	23 + 72 = ______	202 + 410 = ______
391 + 10 = ______	85 – 4 = ______	824 + 73 = ______
981 – 30 = ______	79 – 32 = ______	719 + 230 = ______
620 – 30 = ______	21 + 63 = ______	783 + 12 = ______
459 + 200 = ______	69 – 9 = ______	43 + 544 = ______
59 + 300 = ______	43 + 34 = ______	202 + 177 = ______
394 + 20 = ______	95 – 51 = ______	24 + 635 = ______
290 – 200 = ______	48 – 40 = ______	410 + 544 = ______
618 – 20 = ______	6 + 32 = ______	431 + 30 = ______
117 – 30 = ______	80 + 14 = ______	203 + 280 = ______
202 + 20 = ______	99 – 54 = ______	315 + 50 = ______
1000 – 10 = ______	9 + 40 = ______	300 + 539 = ______

Mental Math 13	Mental Math 14	Mental Math 15
349 – 11 = ______	10 – 3 = ______	9 + 7 = ______
869 – 150 = ______	4 + 5 = ______	90 + 70 = ______
484 – 62 = ______	6 + 7 = ______	39 + 7 = ______
689 – 602 = ______	15 – 8 = ______	390 + 70 = ______
578 – 63 = ______	34 + 23 = ______	139 + 7 = ______
679 – 405 = ______	85 – 3 = ______	391 + 70 = ______
658 – 58 = ______	400 + 200 = ______	139 + 70 = ______
927 – 715 = ______	900 – 300 = ______	60 + 40 = ______
866 – 530 = ______	400 + 50 = ______	60 + 50 = ______
748 – 332 = ______	999 – 20 = ______	60 + 140 = ______
559 – 103 = ______	459 – 8 = ______	60 + 150 = ______
853 – 723 = ______	201 + 4 = ______	60 + 152 = ______
897 – 87 = ______	564 – 40 = ______	348 + 2 = ______
537 – 301 = ______	923 + 70 = ______	348 + 20 = ______
692 – 190 = ______	432 + 34 = ______	570 + 30 = ______
295 – 244 = ______	863 – 42 = ______	580 + 30 = ______
816 – 402 = ______	42 + 127 = ______	580 + 34 = ______
749 – 46 = ______	346 + 643 = ______	582 + 34 = ______
703 – 201 = ______	569 – 237 = ______	269 + 2 = ______
937 – 24 = ______	7 + 462 = ______	269 + 20 = ______

Mental Math 16	Mental Math 17	Mental Math 18
14 – 5 = ______	542 – 9 = ______	461 + 300 = ______
94 – 5 = ______	674 – 7 = ______	394 – 50 = ______
394 – 5 = ______	52 – 6 = ______	490 + 2 = ______
940 – 50 = ______	450 – 60 = ______	978 – 100 = ______
16 – 9 = ______	78 – 2 = ______	342 – 4 = ______
36 – 9 = ______	300 – 90 = ______	5 + 378 = ______
236 – 9 = ______	410 – 50 = ______	200 + 459 = ______
360 – 90 = ______	72 – 8 = ______	821 – 9 = ______
100 – 60 = ______	350 – 80 = ______	891 – 20 = ______
700 – 60 = ______	236 – 7 = ______	30 + 29 = ______
43 – 6 = ______	891 – 8 = ______	8 + 64 = ______
430 – 60 = ______	960 – 20 = ______	110 – 9 = ______
430 – 6 = ______	502 – 50 = ______	110 – 90 = ______
40 – 8 = ______	64 – 8 = ______	800 + 200 = ______
43 – 8 = ______	456 – 9 = ______	705 – 400 = ______
343 – 8 = ______	730 – 50 = ______	6 + 378 = ______
400 – 80 = ______	75 – 9 = ______	60 + 378 = ______
430 – 80 = ______	97 – 1 = ______	600 + 378 = ______
438 – 80 = ______	370 – 80 = ______	745 – 40 = ______
438 – 8 = ______	821 – 50 = ______	82 + 816 = ______

Mental Math 19	Mental Math 20	Mental Math 21
2 x 4 = ______	2 x 4 = ______	2 x 3 = ______
2 x 3 = ______	10 x 2 = ______	6 x 2 = ______
2 x 8 = ______	8 x 2 = ______	1 x 2 = ______
2 x 10 = ______	2 x 1 = ______	2 x 6 = ______
2 x 1 = ______	6 x 2 = ______	5 x 2 = ______
2 x 9 = ______	2 x 7 = ______	2 x 1 = ______
2 x 2 = ______	4 x 2 = ______	2 x 2 = ______
2 x 6 = ______	2 x 6 = ______	8 x 2 = ______
2 x 7 = ______	2 x 9 = ______	2 x 7 = ______
2 x 5 = ______	9 x 2 = ______	2 x 3 = ______
2 x 10 = ______	7 x 2 = ______	2 x 4 = ______
2 x 8 = ______	2 x 2 = ______	2 x 10 = ______
2 x 5 = ______	3 x 2 = ______	7 x 2 = ______
2 x 9 = ______	2 x 5 = ______	2 x 12 = ______
2 x 6 = ______	2 x 10 = ______	4 x 2 = ______
2 x 7 = ______	5 x 2 = ______	10 x 2 = ______
2 x 3 = ______	2 x 3 = ______	8 x 2 = ______
2 x 4 = ______	2 x 2 = ______	2 x 11 = ______
2 x 12 = ______	2 x 8 = ______	9 x 2 = ______
2 x 11 = ______	1 x 2 = ______	2 x 5 = ______

Mental Math 22	Mental Math 23	Mental Math 24
3 x 7 = ______	3 x 3 = ______	3 x 4 = ______
3 x 5 = ______	3 x 2 = ______	8 x 3 = ______
3 x 10 = ______	6 x 3 = ______	5 x 3 = ______
3 x 8 = ______	1 x 3 = ______	3 x 1 = ______
3 x 5 = ______	3 x 6 = ______	2 x 3 = ______
3 x 9 = ______	5 x 3 = ______	9 x 3 = ______
3 x 6 = ______	3 x 1 = ______	3 x 6 = ______
3 x 7 = ______	2 x 3 = ______	7 x 3 = ______
3 x 3 = ______	8 x 3 = ______	10 x 3 = ______
3 x 4 = ______	3 x 7 = ______	3 x 9 = ______
3 x 3 = ______	3 x 3 = ______	3 x 10 = ______
3 x 8 = ______	3 x 4 = ______	3 x 3 = ______
3 x 10 = ______	3 x 10 = ______	4 x 3 = ______
3 x 1 = ______	7 x 3 = ______	11 x 3 = ______
3 x 9 = ______	3 x 9 = ______	3 x 8 = ______
3 x 2 = ______	10 x 3 = ______	3 x 5 = ______
3 x 6 = ______	8 x 3 = ______	3 x 7 = ______
3 x 4 = ______	3 x 8 = ______	12 x 3 = ______
3 x 11 = ______	4 x 3 = ______	3 x 2 = ______
3 x 12 = ______	3 x 5 = ______	6 x 3 = ______

Mental Math 25	Mental Math 26	Mental Math 27
2 x 6 = ______	3 x 4 = ______	2 x 8 = ______
9 x 2 = ______	2 x 1 = ______	12 – 3 = ______
3 x 3 = ______	10 x 3 = ______	9 x 2 = ______
2 x 3 = ______	2 x 9 = ______	2 + 10 = ______
2 x 8 = ______	1 x 3 = ______	3 x 11 = ______
3 x 5 = ______	5 x 2 = ______	11 – 2 = ______
2 x 7 = ______	3 x 7 = ______	8 x 3 = ______
9 x 3 = ______	6 x 2 = ______	7 + 3 = ______
10 x 2 = ______	8 x 3 = ______	2 + 6 = ______
3 x 8 = ______	2 x 2 = ______	3 + 9 = ______
2 x 2 = ______	3 x 9 = ______	1 + 3 = ______
7 x 3 = ______	2 x 10 = ______	9 + 2 = ______
2 x 5 = ______	8 x 2 = ______	8 – 3 = ______
6 x 3 = ______	3 x 2 = ______	12 x 3 = ______
4 x 2 = ______	3 x 3 = ______	8 – 2 = ______
3 x 10 = ______	2 x 3 = ______	2 x 11 = ______
3 x 2 = ______	3 x 6 = ______	3 + 8 = ______
4 x 3 = ______	2 x 4 = ______	3 x 9 = ______
1 x 2 = ______	5 x 3 = ______	12 x 2 = ______
3 x 1 = ______	7 x 2 = ______	10 + 3 = ______

Mental Math 28	Mental Math 29	Mental Math 30
10 ÷ 2 = ______	8 ÷ 2 = ______	14 ÷ 2 = ______
14 ÷ 2 = ______	8 ÷ 1 = ______	2 x 6 = ______
8 ÷ 2 = ______	18 ÷ 2 = ______	16 ÷ 2 = ______
18 ÷ 2 = ______	10 ÷ 1 = ______	7 x 2 = ______
6 ÷ 2 = ______	20 ÷ 2 = ______	18 ÷ 2 = ______
2 ÷ 2 = ______	4 ÷ 1 = ______	6 ÷ 2 = ______
20 ÷ 2 = ______	16 ÷ 2 = ______	2 x 12 = ______
4 ÷ 2 = ______	11 ÷ 1 = ______	10 x 2 = ______
16 ÷ 2 = ______	1 ÷ 1 = ______	20 ÷ 2 = ______
12 ÷ 2 = ______	12 ÷ 2 = ______	8 ÷ 2 = ______
10 ÷ 2 = ______	22 ÷ 2 = ______	8 x 2 = ______
16 ÷ 2 = ______	6 ÷ 1 = ______	2 x 2 = ______
2 ÷ 2 = ______	14 ÷ 2 = ______	12 ÷ 2 = ______
18 ÷ 2 = ______	24 ÷ 2 = ______	10 ÷ 2 = ______
4 ÷ 2 = ______	10 ÷ 2 = ______	11 x 2 = ______
6 ÷ 2 = ______	16 ÷ 1 = ______	4 ÷ 2 = ______
20 ÷ 2 = ______	4 ÷ 2 = ______	4 x 2 = ______
14 ÷ 2 = ______	40 ÷ 2 = ______	2 ÷ 2 = ______
12 ÷ 2 = ______	6 ÷ 2 = ______	2 x 1 = ______
8 ÷ 2 = ______	60 ÷ 2 = ______	2 x 9 = ______

Mental Math 31	Mental Math 32	Mental Math 33
15 ÷ 3 = ______	18 ÷ 3 = ______	12 ÷ 3 = ______
24 ÷ 3 = ______	18 ÷ 2 = ______	9 x 2 = ______
9 ÷ 3 = ______	18 ÷ 1 = ______	10 ÷ 2 = ______
27 ÷ 3 = ______	27 ÷ 3 = ______	6 x 3 = ______
6 ÷ 3 = ______	9 ÷ 3 = ______	12 ÷ 2 = ______
18 ÷ 3 = ______	12 ÷ 2 = ______	7 x 3 = ______
3 ÷ 3 = ______	2 ÷ 2 = ______	27 ÷ 3 = ______
21 ÷ 3 = ______	24 ÷ 3 = ______	12 ÷ 3 = ______
18 ÷ 3 = ______	36 ÷ 3 = ______	12 x 2 = ______
9 ÷ 3 = ______	1 ÷ 1 = ______	9 ÷ 3 = ______
30 ÷ 3 = ______	21 ÷ 3 = ______	8 x 3 = ______
12 ÷ 3 = ______	22 ÷ 2 = ______	2 ÷ 2 = ______
27 ÷ 3 = ______	4 ÷ 2 = ______	10 x 2 = ______
15 ÷ 3 = ______	12 ÷ 3 = ______	16 ÷ 2 = ______
21 ÷ 3 = ______	33 ÷ 3 = ______	12 x 3 = ______
24 ÷ 3 = ______	12 ÷ 1 = ______	18 ÷ 3 = ______
6 ÷ 3 = ______	15 ÷ 3 = ______	7 x 2 = ______
30 ÷ 3 = ______	24 ÷ 2 = ______	6 ÷ 3 = ______
3 ÷ 3 = ______	3 ÷ 3 = ______	10 x 3 = ______
12 ÷ 3 = ______	14 ÷ 2 = ______	9 x 2 = ______

Mental Math 34	Mental Math 35	Mental Math 36
12 ÷ 2 = ______	3 – 3 = ______	9 + 7 = ______
9 x 2 = ______	27 ÷ 3 = ______	13 – 6 = ______
15 ÷ 3 = ______	3 + 3 = ______	123 + 321 = ______
8 ÷ 2 = ______	18 + 2 = ______	8 x 3 = ______
6 x 3 = ______	12 ÷ 2 = ______	38 + 5 = ______
9 ÷ 3 = ______	27 – 3 = ______	2 x 9 = ______
2 x 7 = ______	18 – 2 = ______	81 – 7 = ______
18 ÷ 2 = ______	12 – 3 = ______	348 – 5 = ______
2 ÷ 2 = ______	27 + 3 = ______	235 + 7 = ______
30 ÷ 3 = ______	30 – 3 = ______	999 – 123 = ______
3 x 8 = ______	12 + 3 = ______	3 x 5 = ______
14 ÷ 2 = ______	18 ÷ 2 = ______	27 ÷ 3 = ______
3 x 4 = ______	3 ÷ 3 = ______	40 + 80 = ______
5 x 2 = ______	100 ÷ 100 = ______	7 x 2 = ______
4 ÷ 2 = ______	3 x 1 = ______	120 – 90 = ______
24 ÷ 3 = ______	100 x 1 = ______	18 ÷ 3 = ______
2 x 8 = ______	3 ÷ 1 = ______	683 + 40 = ______
3 ÷ 3 = ______	25 ÷ 1 = ______	18 ÷ 2 = ______
3 x 9 = ______	100 ÷ 1 = ______	683 – 40 = ______
21 ÷ 3 = ______	100 – 1 = ______	99 + 9 = ______

Place-Value Cards

1	10	100
2	20	200
3	30	300
4	40	400
5	50	500

Place-Value Cards

6	60	600
7	70	700
8	80	800
9	90	900
1000		

Hundred Chart

1	2	3	4	5	6	7	8	9	10
11	12	13	14	15	16	17	18	19	20
21	22	23	24	25	26	27	28	29	30
31	32	33	34	35	36	37	38	39	40
41	42	43	44	45	46	47	48	49	50
51	52	53	54	55	56	57	58	59	60
61	62	63	64	65	66	67	68	69	70
71	72	73	74	75	76	77	78	79	80
81	82	83	84	85	86	87	88	89	90
91	92	93	94	95	96	97	98	99	100

2.1a

1. Harriet has 9 parakeets and 5 parrots. How many birds does she have?

2. The noise from the birds was just too much and so she gave 8 birds away. How many birds does she have left?

3. A friend gave her 2 macaws. How many birds does she have now?

4. Her landlord complained about the noise and she had to give some birds away again. Now she just has 2 quiet parakeets left. How many birds did she have to give away the second time?

2.1b

1. Harriet started out with 9 parakeets and 5 parrots. How many more parakeets than parrots did she have?

2. After giving some birds away, Harriet ended up with 2 parakeets. She was finally able to move out of an apartment and get a large house. She now has 20 birds. How many more birds does she have now than she had before?

3. Harriet saw a blue parakeet at the pet store. It cost $18. She had $6 with her. How much more money does she need?

2.1d

1. Harriet now has 20 birds. Some of them laid eggs. There were 37 eggs altogether. 25 of the eggs hatched.

   (a) How many birds does she now have?

   (b) How many eggs did not hatch?

   (c) She sold 13 of her birds. How many does she have left?

   (d) How many more birds does she have now than she had before they started laying eggs and she sold some?

2.5d

1. Harriet bought 148 bags of birdseed. So far she has used 92 bags. How many bags does she have left?

2. Harriet saw a cockatoo for sale at a breeder's. It cost $750. She had $680. How much more money does she need to buy the cockatoo?

3.1b

1. Harriet needed to make some cages for her birds. She wanted to make two cages. For one, she needed 48 meters of chicken wire. For the other, she needed 76 meters of chicken wire.
   (a) How many meters did she need in all?

   (b) She bought 200 meters of wire. How many extra meters does she have?

2. August lives 304 meters from school. May lives 283 meters from school. Who lives closer to the school? How much closer to the school does that person live?

3. Three trees are in a line. The first tree is 105 m from the third tree. The second tree is 63 m from the third tree. How far is the second tree from the first tree?

4. A skyscraper is 120 meters tall. A house is 104 meters shorter than the skyscraper. An apartment building is 38 meters taller than the house.
   (a) How tall is the house?
   (b) How tall is the apartment building?
   (c) If the house were put on top of the apartment, would it be as high as the skyscraper? How much higher or lower is the skyscraper than the apartment building and house together?

5. There are three ropes, A, B, and C. The total length of the ropes is 94 meters. Rope A is 51 meters long. Rope C is 38 meters long. How long is Rope B?

Which is Longest?

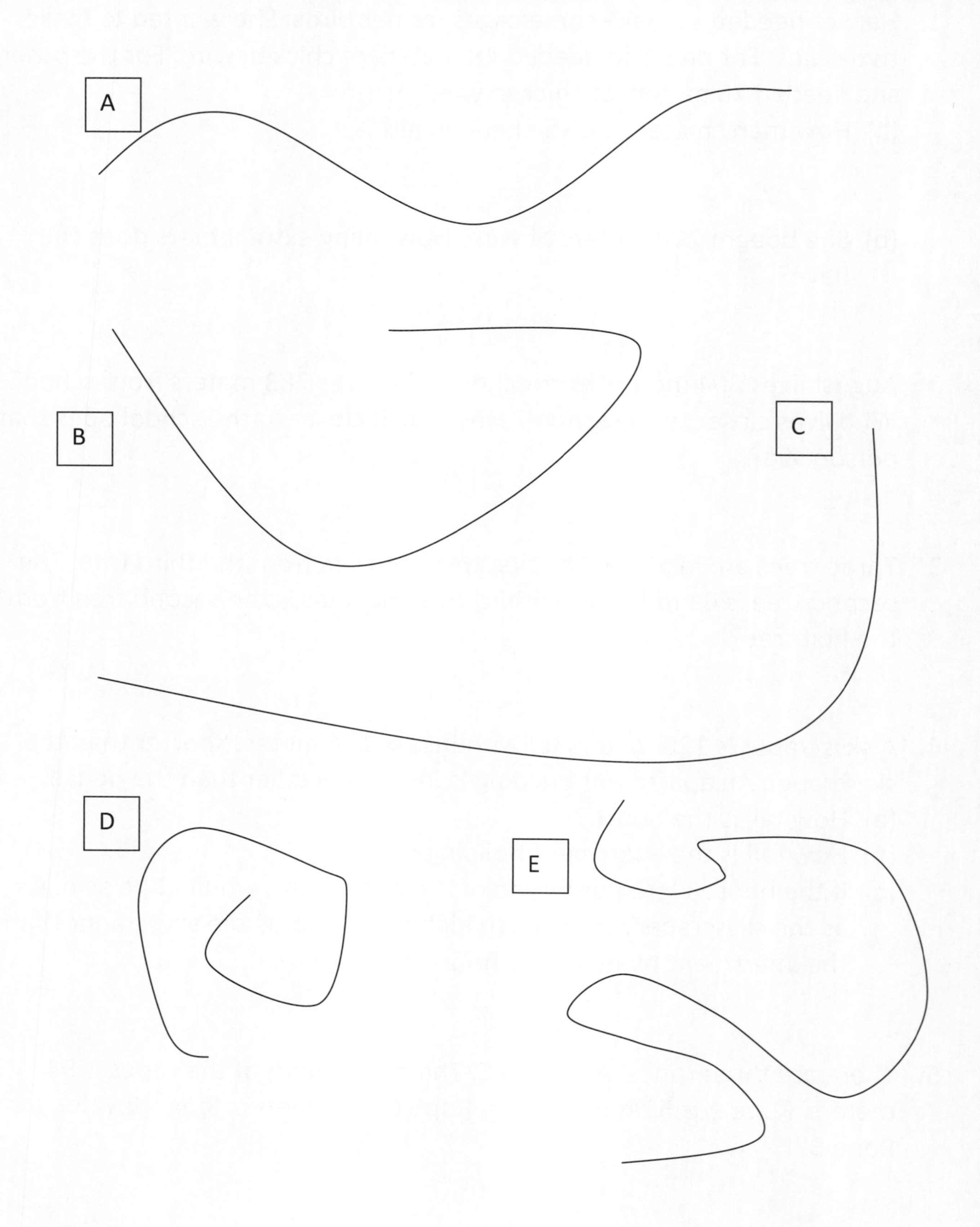

1. A snail is crawling up one of the posts Harriet made for her birds to perch on. The post is 39 inches tall and is slippery. If the snail crawls up the post 15 inches every hour, but slides back 7 inches when he rests at the end of each hour, how many hours will it take the snail to get to the top?

2. There are 4 sticks.
   Stick D is between Stick A and B.
   Stick C is the shortest.
   Stick B is shorter than Stick A.
   Arrange the sticks in order, starting with the shortest stick.

4.1a

1. Box A is heavier than Box B. Box B weighs the same as Box C. Box C is heavier than Box D. Which is lightest, A, B, C, or D?

2. Box X and 2 kg weigh the same as Box Y and 3 kg. How much heavier is Box X than Box Y?

Number Lines

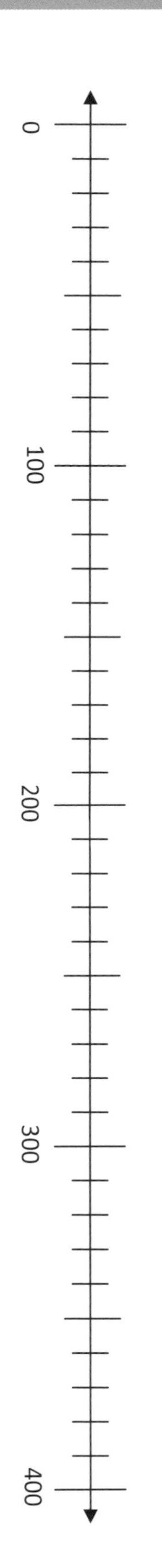

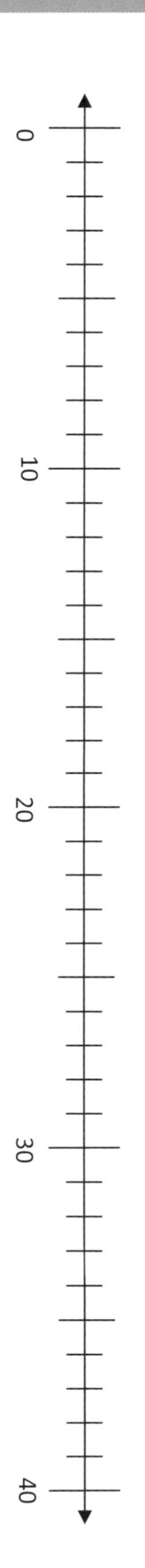

Circle Scale

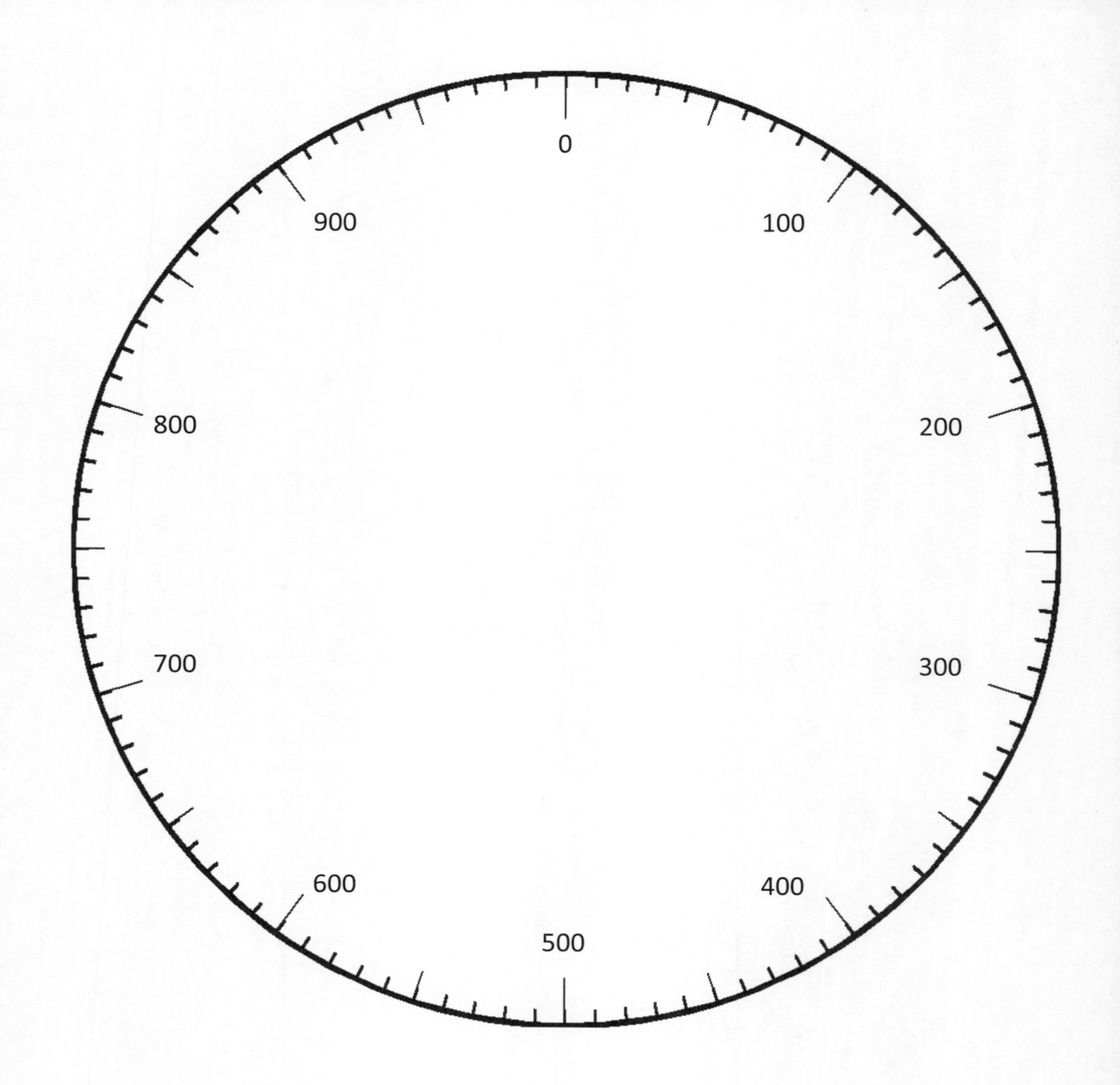

Pounds and Ounces

1. Fill in the blanks with **oz** or **lb**.

   (a) 40 pennies weigh about 4 ________.

   (b) A cat weighs about 8 ________.

   (c) A bee weighs about 1 ________.

   (d) A banana weighs about 5 _________ .

   (e) A bag of flour weighs 3 _________.

2. A bag of beans weighs about 10 pounds. It also weighs about (circle the best answer):

   12 kilograms   20 ounces   200 ounces   200 grams

3. A book weighs 23 ounces. It is put on one side of the balance and some weights are put on the other side of the balance. The weights weigh 8 ounces, 4 ounces, or 1 ounce. There are 4 of each type of weight. How many of each can you use to balance the book?

   8 oz ________

   4 oz ________

   1 oz ________

4. A cook used 35 pounds of potatoes. There were 18 pounds left. How many pounds of potatoes were there at first?

5. Joyce weighs 64 pounds. She weighs 16 pounds more than Jorge. How much does Jorge weigh?

1. Harriet bought some seed cakes for her birds. They ate 25 seed cakes in 5 days. How many seed cakes did they eat in one day?

2. Harriet had 3 friends over to see her birds. She made 24 cookies. She and her friends shared the cookies equally. How many cookies did Harriet get?

1. One night, Harriet wanted find out how many birds were on a perch without shining the flashlight into their faces. She counted 30 feet. How many birds were on the perch?

2. Harriet wants to take 24 birds to a bird show. She wants to put 3 birds in each cage. How many cages does she need?

3. If she wants to take 26 birds instead, how many cages would she need? For 25 birds, how many cages would she need?

1. Harriet has 16 cookies and wants to divide them up among herself and three friends. How many would each person get?

2. Harriet has 20 cookies. She wants to put 4 cookies on each plate. How many plates does she need?

3. There are 6 plates of cookies with 5 cookies on each plate. How many cookies are there?

4. On a rainy day, Harriet's friends all wore boots when they came to visit. They left 12 boots by the door. How many friends came to visit?

5. You see some dogs and count 32 legs. How many dogs are there?

6. Harriet uses 14 jugs of water to give water to her birds in a week. If she used the same number of jugs each day, how many jugs did she use each day?

7. A string 12 inches long is cut into equal parts, each 3 inches long. How many equal parts are there?

8. A string 12 inches long is cut into two equal parts. How long is each part?

9. There are 3 feet in a yard and 12 inches in a foot. How many inches are there in a yard?

10. A decimeter is 10 centimeters. How many decimeters are there in a meter?

Multiplication Table for 2

2 x 1 = ______	1 x 2 = ______
2 x 2 = ______	2 x 2 = ______
2 x 3 = ______	3 x 2 = ______
2 x 4 = ______	4 x 2 = ______
2 x 5 = ______	5 x 2 = ______
2 x 6 = ______	6 x 2 = ______
2 x 7 = ______	7 x 2 = ______
2 x 8 = ______	8 x 2 = ______
2 x 9 = ______	9 x 2 = ______
2 x 10 = ______	10 x 2 = ______
2 x 11 = ______	11 x 2 = ______
2 x 12 = ______	12 x 2 = ______

Multiplication Chart

x	1	2	3	4	5	6	7	8	9	10
1										
2										
3										
4										
5										
6										
7										
8										
9										
10										

Multiplication Charts

x	5	10	2	1	8	9	3	7	4	6
1										
2										

x	10	4	5	2	7	3	8	9	6	1
1										
2										

x	3	6	1	9	5	4	7	10	8	2
2										
1										

Multiplication Chart

x	1	2	3	4	5	6	7	8	9	10
1										
2										
3										
4										
5										
6										
7										
8										
9										
10										

Multiplication Table for 3

3 x 1 = ______	1 x 3 = ______
3 x 2 = ______	2 x 3 = ______
3 x 3 = ______	3 x 3 = ______
3 x 4 = ______	4 x 3 = ______
3 x 5 = ______	5 x 3 = ______
3 x 6 = ______	6 x 3 = ______
3 x 7 = ______	7 x 3 = ______
3 x 8 = ______	8 x 3 = ______
3 x 9 = ______	9 x 3 = ______
3 x 10 = ______	10 x 3 = ______
3 x 11 = ______	11 x 3 = ______
3 x 12 = ______	12 x 3 = ______

Multiplication Charts

x	5	10	2	1	8	9	3	7	4	6
2										
3										

x	3	6	1	9	5	4	7	10	8	2
2										
3										

x	10	4	5	2	7	3	8	9	6	1
3										
2										

Multiplication Game Board

24	15	14	2	27	30
21	18	9	16	4	12
4	24	6	8	12	12
12	30	27	9	20	15
20	2	18	16	3	6
6	2	8	3	27	14
10	6	21	18	18	10

______ x 2 = 2	2 ÷ 2 = ______
______ x 2 = 4	4 ÷ 2 = ______
______ x 2 = 6	6 ÷ 2 = ______
______ x 2 = 8	8 ÷ 2 = ______
______ x 2 = 10	10 ÷ 2 = ______
______ x 2 = 12	12 ÷ 2 = ______
______ x 2 = 14	14 ÷ 2 = ______
______ x 2 = 16	16 ÷ 2 = ______
______ x 2 = 18	18 ÷ 2 = ______
______ x 2 = 20	20 ÷ 2 = ______
______ x 2 = 22	22 ÷ 2 = ______
______ x 2 = 24	24 ÷ 2 = ______

6.3b

1. There are 16 posts with branches in one of Harriet's bird cages. 2 birds are sitting on each post. How many posts are there?

2. Harriet took 10 birds to a show. She wants to put 2 birds in each cage. How many cages does she need?

3. For another show, she took 7 birds. She again wants to put 2 birds in each cage. How many cages does she need?

1. Harriet bought 18 pounds of birdseed. The birdseed came in 2 bags. How many pounds did each bag weigh?

2. Harriet used 6 pounds of the birdseed. How many pounds are left?

3. For the remaining birdseed, she used 2 pounds a day. How many days did it take to use up the birdseed?

4. She needed to buy more birdseed. This time she bought 18 pounds and an additional 7 pounds of birdseed. How many pounds of birdseed does she have now?

5. Harriet bought two new birds. One cost $398 and the other cost $276. How much did they both cost?

6. Harriet had $800 to spend on the birds. How much money did she have left over?

Division Table for 3

1 x 3 = ______ | ______ ÷ 3 = 1

2 x 3 = ______ | ______ ÷ 3 = 2

3 x 3 = ______ | ______ ÷ 3 = 3

4 x 3 = ______ | ______ ÷ 3 = 4

5 x 3 = ______ | ______ ÷ 3 = 5

6 x 3 = ______ | ______ ÷ 3 = 6

7 x 3 = ______ | ______ ÷ 3 = 7

8 x 3 = ______ | ______ ÷ 3 = 8

9 x 3 = ______ | ______ ÷ 3 = 9

10 x 3 = ______ | ______ ÷ 3 = 10

11 x 3 = ______ | ______ ÷ 3 = 11

12 x 3 = ______ | ______ ÷ 3 = 12

Division Game Board

1	4	9	3	5	☆
1	4	6	7	5	6
2	3	8	7	4	6
10	7	5	10	8	6
2	9	8	3	1	10
9	5	8	1	3	4
☆	2	10	2	8	7

Hundreds	Tens	Ones

Blank Page